From Molecules to Crystallizers

Roger J. Davey & John Garside

Department of Chemical Engineering
UMIST
Manchester
U.K.

Series sponsor: **ZENECA**

ZENECA is a major international company active in four main areas of business: Pharmaceuticals, Agrochemicals and Seeds, Speciality Chemicals, and Biological Products.

ZENECA's skill and innovative ideas in organic chemistry and bioscience create products and services which improve the world's health, nutrition, environment, and quality of life.

ZENECA is committed to the support of education in chemistry and chemical engineering.

OXFORD
UNIVERSITY PRESS

OXFORD

UNIVERSITY PRESS

Great Clarendon Street, Oxford OX2 6DP

Oxford University Press is a department of the University of Oxford.
It furthers the University's objective of excellence in research, scholarship,
and education by publishing worldwide in

Oxford   New York

Athens   Auckland   Bangkok   Bogotá   Buenos Aires   Calcutta
Cape Town   Chennai   Dar es Salaam   Delhi   Florence   Hong Kong   Istanbul
Karachi   Kuala Lumpur   Madrid   Melbourne   Mexico City   Mumbai
Nairobi   Paris   Saõ Paulo   Singapore   Taipei   Tokyo   Toronto   Warsaw

with associated companies in   Berlin   Ibadan

Oxford is a registered trade mark of Oxford University Press
in the UK and in certain other countries

Published in the United States
by Oxford University Press Inc., New York

A catalogue record for this book is available from the British Library

Library of Congress Cataloging in Publication Data
(Data applied for)
ISBN 0 19 850489 6
Typeset by EXPO Holdings, Malaysia
Printed in Great Britain
on acid-free paper by Bath Press Ltd., Bath, Avon

Series Editor's Foreword

Oxford Chemistry Primers have been designed to provide concise introductions to topics commonly encountered in chemistry, and more recently, chemical engineering undergraduate courses. In the context of chemical engineering no such series would be complete without a Primer on crystallization. Quite apart from its industrial importance, crystallization provides an excellent example of how the skills and knowledge base of the chemist, materials scientist and chemical engineer can be brought together to design a product with the particular attributes required for its application. The shear diversity of products containing components that have been manufactured in crystalline form—including, for example, cosmetics, food and pharmaceuticals—also hints at the wide range in chemical knowledge required in this field and the importance of being able to produce such products to very exacting specification.

In this Primer, two international leaders in the field of crystallization—Roger Davey and John Garside—have worked together to produce a text which has drawn together the relevant chemistry and chemical engineering to show how, in the real world, molecular-scale science and chemical engineering process design are integrated to produce the crystalline materials we require in our everyday lives. This is a book that can be understood by both chemistry and chemical engineering undergraduates—I hope both communities will read, learn and enjoy!

Lynn F. Gladden
Department of Chemical Engineering
University of Cambridge

Preface

Crystalline materials form an essential part of our modern technological environment. From metals and ceramics, cements and plastics through pharmaceuticals, foods, cosmetics, pigments and a myriad of electronic components, we use and rely on crystallinity every day of our lives. While crystallization and crystallography are essentially branches of physical chemistry, the production of crystalline materials relies additionally on chemical engineering for the design and operation of appropriate processes and equipment. In this book we have attempted to provide a multidisciplinary perspective of the processes by which molecules may be transformed into crystals and hence into products. We have combined crystallography, thermodynamics, chemical kinetics, reactor modelling and surface chemistry in a format which we hope will demonstrate how all these elements are integral to the formation of crystalline materials. Throughout we have attempted to give the text an historical perspective and to illustrate theoretical concepts by use of data on familiar materials. In this way we hope to have linked the fundamental aspects of molecular self-assembly through crystal formation to crystallizer design and product formulation, in a comprehensive and accessible way.

We would particularly like to thank Harvey Alison, Nicholas Blagden, Mike Quayle, Linda Seton-Williams and Elizabeth Worthington for their help with the preparation of the diagrams and the text. Most especially we thank Trish and Pat for their continued interest, their support and their forbearance.

Manchester
September 1999

R.J.D.
J.G.

Contents

1 Introduction

1.1 Why crystallization?

Crystallization is a purification technique, a separation process and a branch of particle technology. It thus encompasses key areas of chemical and process engineering. It is a supramolecular process by which an ensemble of randomly organized molecules, ions or atoms in a fluid come together to form an ordered three-dimensional molecular array which we call a crystal. The formation and growth of crystals is thus also of fundamental scientific interest.

The power of crystallization arises from the reproducibility of this assembly process and the nature of crystals themselves. Because the constituent atoms, ions or molecules of a crystal are arranged in a regular infinite manner, crystal surfaces are highly discriminating. In general, they only allow similar molecular-scale growth units to attach themselves to the crystal lattice. As a result crystals are highly effective in their separating power.

When a crystal forms, something like 10^{20} molecules come together in an array which is identical no matter where in the world the process takes place.

Many commonly encountered solid materials are crystalline—sugar and salt, diamond and quartz, icicles and snowflakes. Other common materials may not be immediately recognizable as crystalline either because of their small size or, more often perhaps, because they form part of a composite material. Thus, cement contains crystalline calcium hydroxide and silicate; paints contain pigment crystals; most pharmaceutical materials are prepared in the crystalline state and subsequently ground to yield powders; washing powders are mixtures of crystalline carbonates, phosphates and perborates. Crystals that we encounter are not just man-made; eggs and mollusc shells are composed primarily of crystalline calcium carbonate, while bones and teeth are mainly calcium phosphates.

The properties of crystals are often directionally dependent—they are anisotropic. They grow and dissolve at different rates in different directions, their refractive indices vary with direction, and their thermal expansion coefficients and electrical conductivity are directionally dependent, as are their mechanical properties.

Crystals can be almost any size from a few nanometres (active phases in catalysts), through submicrometre (pigments, pharmaceuticals, silver halides in photographic emulsions) up to several millimetres (salt, sugar and diamonds). Size is no indication of crystallinity. The production of crystals having sizes smaller than a few micrometres is traditionally referred to as *precipitation*.

1.2 What do we need to know?

Before we can understand the way in which crystallization processes work, we need to understand a number of different phenomena. An appreciation of the

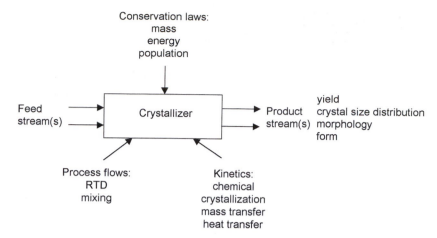

Fig. 1.1   A framework for describing crystallizer behaviour.

underlying phase equilibrium is clearly important; we need to know what crystalline phases it is possible to produce under known conditions—this is a question of thermodynamics. We must also understand issues surrounding the rates at which crystallization takes place—here it is the kinetics of nucleation (the rate at which new crystals are formed) and of growth (the rate at which existing crystals grow) that are of greatest importance. Both of these kinetic events have their basis in molecular-level phenomena and this implies that we also need some understanding of solid state chemistry.

At the process level, control of crystal size and size distribution is important. Here we have to develop an understanding of population balances to set alongside the mass and energy balances that are written instinctively by chemical engineers. Finally, the process flows within the section of process plant where a crystallization is taking place must be known. Is the process batch or continuous? Is the system fully mixed with respect to the solution and the crystals or does the residence time distribution (RTD) of one or both of the phases differ from that corresponding to the fully mixed case? The answers to these questions depend on the nature of the process flows, and hence on the fluid/particle mechanics of the crystal suspension within the crystallizer.

These various features form a framework that will guide the structure of this book (Fig. 1.1). The interactions between the kinetics, the process flows within the crystallizer and the crystal size distribution are illustrated diagrammatically in Fig. 1.2 and the recognition of these interactions was the basis for the first rigorous process engineering description of crystallization (Randolph and Larson, 1988).

1.3   What are crystals and how do we recognize them?

Formally a solid is defined by the way in which its constituent atoms, ions or molecules are packed. If this packing is regular, forming an infinite three-

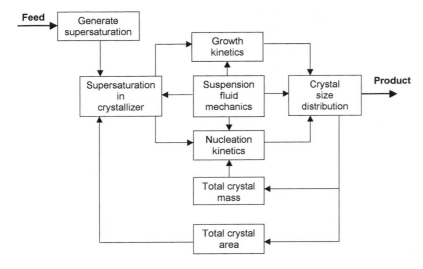

Fig. 1.2 Interactions influencing crystallizer behaviour.

dimensional array, the material is said to be crystalline. There is long-range order. If no such infinite array exists, there is only short-range order and the material is classified as amorphous; glass and rubber are such examples.

The symmetry of the ordered array of molecules in a crystal (the lattice) is expressed by crystallographers as a *space group* which is a combination of symmetry operations that enables a complete crystal structure to be generated from a single (or group of) molecule(s), the asymmetric unit. One important consequence of the ordered packing of molecules in a crystal is that crystals have regular shapes. They grow as polyhedra, never as spheres, with the crystal shape being a characteristic of the material in question.

For example, sodium chloride crystals are cubes, alum crystals are octahedral, and sulphur grows as needles. The symmetry inherent in the shape, or *morphology*, of a crystal reflects the underlying packing symmetry of the crystal lattice and is called the *point group symmetry*.

The molecular order inherent in crystals enables the identification of crystalline materials to be made by a number of specific techniques. Crystals larger than about 5 μm may be imaged directly in the optical microscope. Crystalline particles are often birefringent and hence may be identified by use of crossed polars when they appear as bright areas on a black background. Crystalline solids have well-defined melting points that may be measured to give an indication of crystallinity. As a result of their long-range order, crystalline solids demonstrate a further property which is the most important in characterizing crystallinity—diffraction of X-rays and electrons. Chemists are familiar with using NMR, UV and IR spectroscopy to fingerprint single molecules; diffractograms, such as that in Fig. 1.5, give an equivalent fingerprint for a collection of molecules as they exist in a crystal.

Often it is important to measure the size of crystals that have been produced. This can be done by a number of techniques. Laser diffraction,

Fig. 1.3 The crystal structure of naphthalene. Crystals are held together by combinations of electrostatic and van der Waals forces. Naphthalene is an aromatic hydrocarbon: note the regular, herring-bone pattern of molecules as they are packed in the crystal.

Crystal surfaces are smooth and planar (faceted). Although the relative facial areas may vary, within a sample of a given material the interfacial angles are constant.

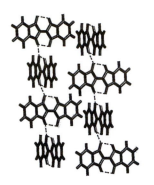

Fig. 1.4 The crystal structure of indigo, the dye used to colour jeans blue. Molecules are held together by hydrogen bonds shown here as broken lines.

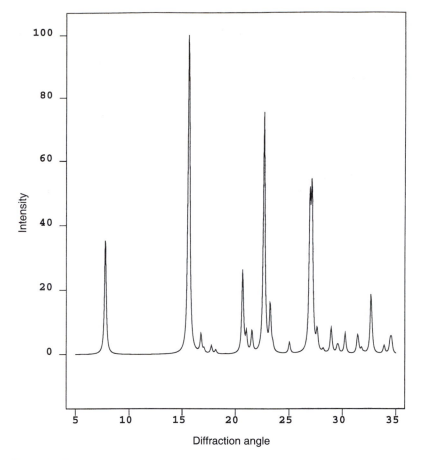

Fig. 1.5 The X-ray diffractogram of the painkiller aspirin. It is expressed in terms of the intensity of diffracted radiation as a function of diffraction angle, 2θ.

sieving and electric zone sensing are commonly used for particles in the size range 1 to 2000 μm. Light scattering and electron microscopy are used for sizes from about 3 μm down to 3 nm.

1.4   Book organization

In this book we explore many of the issues surrounding the conditions under which crystals form in a system and the kinetic processes of nucleation and growth that give rise to crystalline materials. We then describe the factors that control commercially important parameters such as crystal morphology and crystal structure, as well as the development of processes for the control of crystal size. Finally, the importance of crystalline materials in formulated products such as pharmaceuticals and foods is described.

References

Randolph A. D. and Larson M. A. (1988) *Theory of particulate processes*, 2nd edn, Academic Press, New York.

Further reading

For further information on crystallography see:

Bunn C. W. (1945) *Chemical crystallography*, Oxford University Press.
Clegg W. (1998) *Crystal structure determination*, Oxford Chemistry Primer No. 60, Oxford University Press.

A comprehensive reference on particle size analysis is:

Allen T. (1992) *Particle size measurement*, 4th edn, Chapman and Hall, London.

2 Phase equilibria and crystallization techniques

In this chapter we look first at the use of the information available in phase diagrams to define the nature of the crystalline phase that will appear under a given set of conditions of temperature, pressure and composition. We then examine the nature of the driving force for crystallization processes and finally discuss some typical techniques used for preparing crystalline materials.

2.1 The phase rule

A phase is defined as a homogeneous, physically distinct and mechanically separable portion of a system. Examples would be gases, pure liquids (melts), solids and solutions.

Equilibrium is a state of rest of a system. At constant temperature, T, pressure, P, and composition, x, the phases which comprise the equilibrium state will remain constant for infinite time. A change in T, P or x will change the equilibrium state. For a given T, P and x the same state of equilibrium is reached, no matter how it is approached.

As an example consider sodium chloride crystals and water. If excess sodium chloride is shaken with water at 30°C the equilibrium composition of the solution phase will be 36.1 g NaCl per 100 g H_2O. This is the solubility of sodium chloride in water at 30°C. If the composition of the water were changed by adding hydrochloric acid (HCl) at 30°C then the equilibrium NaCl content of the solution would fall to a value that depends on the HCl level in the water. This decrease would be governed by the constancy of the *solubility product*,

$$K_{sp} = a_{Na^+} a_{Cl^-} \qquad (2.1)$$

The solubility of sodium chloride in water is shown in Figs 2.1 and 2.3. The solubility represents the maximum concentration of this salt that can be dissolved at any temperature. If a solution saturated at 40°C were allowed to cool to 30°C then NaCl would crystallize until the solution phase composition reached its equilibrium value, 36.1g NaCl per 100 g H_2O, at this temperature.

where a is the ionic activity.

For a saturated aqueous solution of sodium chloride in contact with solid salt the *number of phases* is three: solid NaCl, aqueous solution and vapour. There are two *constituents* of the system: salt and water. The composition of any phase can be expressed in terms of NaCl and H_2O and hence there are two *components* in the system. The number of components is the minimum number of chemical compounds (constituents) required to express the composition of any phase.

Every system has a *number of degrees of freedom*, which is the number of the variables (temperature, pressure and composition) that must be fixed in order to specify an equilibrium state. In 1876, Gibbs showed that the relationship between the number of phases, P, number of components, C, and degrees of freedom, F, is

$$P + F = C + 2.$$

This is known as the *phase rule*.

The phase rule tells us under which conditions certain systems can be in equilibrium. For example, in a one-component system such as H_2O, the phase rule tells us that for the three phases steam, water and ice to coexist at equilibrium,

$$F = 1 + 2 - 3 = 0.$$

Thus there are no degrees of freedom and so equilibrium can only exist at one temperature and pressure.

If equilibrium is established between a solid and a solution (e.g. NaCl crystals and water) then the solution becomes saturated and there are two components and three phases (solid, liquid and vapour), hence:

$$F = 2 + 2 - 3 = 1.$$

By specifying any one of the variables T, vapour pressure or x(solution), an equilibrium state is defined. If the value of one of these variables is changed, a new equilibrium state will be reached. This is reflected, for example, in the sodium chloride–water solubility curve shown in Fig. 2.1. Thus the phase rule may be used in a qualitative way to describe and define equilibrium states. Measured data on the relation between T, P, x and the crystal structure of solid phases are always required for specific systems and are usually represented on a *phase diagram*.

2.2 Phase diagrams

Solubility curves

Returning to the NaCl/H_2O example it is possible, by experiment, to determine the composition of saturated aqueous solutions in contact with solid salt at any temperature. When these data points are plotted as solubility versus temperature, the solubility curve is produced. For NaCl the solubility in water is almost independent of temperature. For other materials the solubility may be a strongly increasing function of temperature as in the case of sucrose or citric acid. These examples are shown in Fig. 2.1.

Conceptually, creating a solution from a solid and a solvent may be considered in a number of stages:

1. melt the solid;
2. mix the molten solid with solvent;
3. allow to become homogeneous.

In some systems, defined as *ideal*, the only enthalpy change involved in this process is that for melting the solid. Then the solubility, x_{eq} is given by:

$$\ln x_{eq} = \frac{\Delta H_f}{R} \left[\frac{1}{T_f} - \frac{1}{T} \right]. \tag{2.2}$$

It is worth noting that for some elements and compounds it is possible to prepare two or more different crystal structures so that from one molecule we can crystallize at least two different materials. Well-known examples of this phenomenon are graphite and diamond which are different forms of carbon, and calcite, aragonite and vaterite which are all calcium carbonate. This is known as *polymorphism* and is so important that we have devoted the whole of Chapter 6 to this topic.

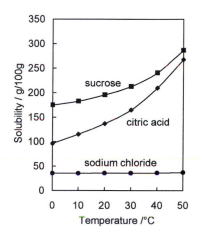

Fig. 2.1 Solubility curves.

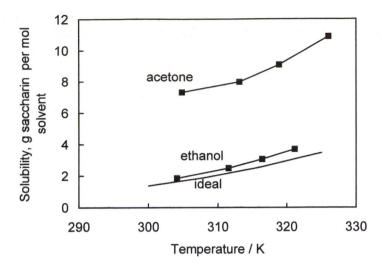

Fig. 2.2 Saccharin is a well-known artificial sweetener. Its solubility in ethanol and acetone is shown here together with its ideal solubility calculated using Eqn 2.2. In ethanol it behaves almost ideally, while in acetone the solubility is greatly enhanced due to solvation.

In other systems the mixing process is accompanied by considerable change in enthalpy due to solvent–solute interactions; such solutions are *non-ideal*.

The variation of solubility with temperature is often described by an equation based on 2.2 but using empirical constants *a*, *b* and *c* fitted from experimental data:

$$\log x_{eq} = a + \frac{b}{T} + c \, \log T. \tag{2.3}$$

A compilation of data in this form for inorganic two-component systems is given by Broul *et al.* (1981). As an alternative an empirical power law equation of the form

$$c_{eq} = A + B\theta + C\theta^2 + \dots \tag{2.4}$$

is often used where c_{eq} is the solubility, θ is the temperature (°C) and A, B, C, etc. are fitted constants.

If in the NaCl/H$_2$O system measurements are made below 0°C, then the equilibrium solution compositions are found to lie on a second solubility curve as shown in Fig. 2.3. Analysis of the solid phase present (by X-ray diffraction for example) would indicate that a hydrate, NaCl.2H$_2$O, has formed. The line DC is the solubility curve of anhydrous NaCl and CB that of the dihydrate. At the point C the solution is saturated with respect to both of the solid constituents. Thus the system of two components exists in four phases—two solid phases, a solution and a vapour—and so

$$F = 2 + 2 - 4 = 0.$$

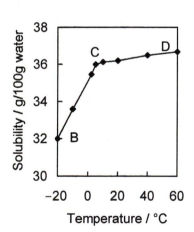

Fig. 2.3 Solubility of NaCl in water.

The system has no degrees of freedom and therefore it is not possible to change any variable and retain equilibrium. By introducing an extra component, say HCl, then a degree of freedom may be introduced so that if the composition is changed the equilibrium point C will move to a different temperature, while if the temperature is changed it will adopt a new composition. It is clear from this type of phase diagram that NaCl may be crystallized at temperatures above 0°C while NaCl·2H$_2$O crystallizes at lower temperatures. The point C is called a *transition point*.

Finally we note, again using the sodium chloride example, that solubility may also be varied by adding a third component in the form of a miscible cosolvent. Now the phase rule tells us that such a system (NaCl, water and, say, acetone) has two degrees of freedom. The equilibrium position could be changed either by a change in temperature or composition. It is common for the crystallization of organic molecules to be achieved by such a reduction in solubility, a process often referred to as *drowning out*.

Eutectic systems

The solubility curves of Figs 2.1 to 2.3 are regions of much larger phase diagrams in which the composition of the two components varies continuously from pure 'solvent' to pure 'solute'. In the case of solubility the term solvent is used to describe the component which is in excess. Such a distinction between solvent and solute becomes meaningless in the extended diagram; instead we shall refer to components A and B with melting points T_A and T_B with $T_B > T_A$.

A phase diagram may be constructed for this system by making a range of two-component (A and B) mixtures and measuring either their melting temperatures or the temperatures at which B just dissolves or at which A just crystallizes.

Figure 2.5 is a generic diagram of this sort. The line BC corresponds to solutions of B in A while the line AC corresponds to solutions of A in B. Their point of intersection, C, is the *eutectic* point. Along the line AC pure A is in equilibrium with A/B liquid, along BC pure B is in equilibrium with A/B liquid, while at point C solid A, solid B, liquid of composition C and vapour are in equilibrium. This means that at C there are no degrees of freedom and hence the eutectic temperature and composition are fixed. In a three-component system the eutectic temperature varies with composition.

The line ACB is referred to as the liquidus since it gives the equilibrium composition of the liquid phase. The line ADCEB is the solidus since it gives the composition of the corresponding solid phase. Thus, for example, the point M represents a mixture of solid A (composition S) and liquid of composition L. Examples of this type of system are gold/thallium, naphthalene/benzene, *meta*- and *para*- nitrochlorobenzene and KCl/H$_2$O.

From the point of view of crystallization it is important to appreciate that in a eutectic system, while it is possible to crystallize pure components from an impure mixture in a single step, the maximum yield achievable is limited by the existence of the eutectic point.

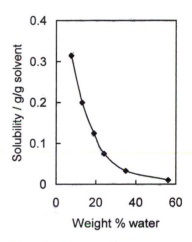

Fig. 2.4 The explosive, pentaerythritol tetranitrate is crystallized by the addition of acetone to an aqueous solution. Here is its solubility curve at 25°C.

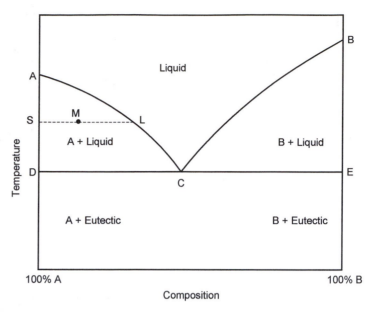

Fig. 2.5   Eutectic phase diagram.

Compounds and solid solutions

In some systems the two components can co-crystallize either as a *compound* as in the case of water and salt in $NaCl.2H_2O$ or as a *solid solution* as with NaCl and NaBr or naphthalene and β-naphthol. In such systems it is never possible to crystallize pure components in a single step. The reason for this is clear from Fig. 2.6, which is a typical phase diagram for such a two-component mixture.

Consider the liquid of composition M. On cooling along MN the point L is reached at which solid of composition S is precipitated. Further cooling creates more solid solution, whose composition changes progressively along SS', in contact with liquids of compositions lying along LL'. The last solid to form has composition S' and the remaining liquid a composition L'.

The importance of phase diagrams

The guidance of the phase rule and the availability of phase diagrams are important adjuncts to the development of crystallization processes. They provide vital data on the possible phases crystallizing, the likely process yields and the number of variables that may be utilized to control a process. Where no data are available for a specific product, the measurement of solubility and heats and temperatures of melting will be an important first phase of development work.

2.3   The driving force for crystallization

Crystallization is a kinetic process and traditionally chemists are familiar with the use of reagent concentration to describe the driving force for the chemical

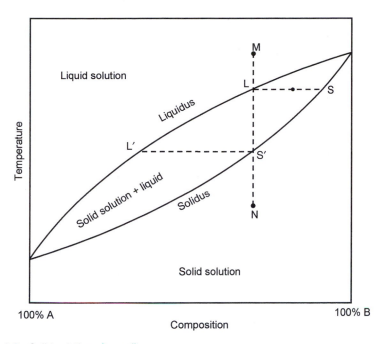

Fig. 2.6 Solid solution phase diagram.

rate processes. However, in the case of crystallization the concentration range over which the process can occur is limited by the equilibrium composition of the system corresponding to the conditions chosen.

Figure 2.7 shows a hypothetical solubility curve. A solution whose composition lies below the solubility curve is *undersaturated* and existing crystals will dissolve. A solution lying above the solubility curve is termed *supersaturated*, since the amount of dissolved solute is greater than the equilibrium saturation value. Crystals can nucleate and grow only if the

In the 19th century the Nobel prize-winning physical chemist Wilhelm Ostwald was a major investigator of crystallization phenomena. He discovered that at comparatively low supersaturations, although existing crystals will grow, it is difficult to create new crystals. Once some critical level of supersaturation is exceeded, new crystals form spontaneously and the solution is now *labile*. The region (see Fig. 2.7) between equilibrium and labile states is called the *metastable zone*. Its width depends on the ease of production of new crystals and so is determined by kinetic rather than thermodynamic factors.

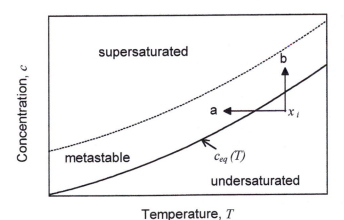

Fig. 2.7 The solubility/supersolubility diagram.

solution is supersaturated and so the production of a supersaturated solution is a prerequisite for crystallization. It is this supersaturation, expressed by comparing the actual to the equilibrium composition, which drives the crystallization process.

Formally the supersaturation, σ, can be defined in thermodynamic terms as the dimensionless difference in chemical potential between a molecule in an equilibrium state, μ_{eq}, and a molecule in its supersaturated state, μ_{ss}:

$$\sigma = (\mu_{ss} - \mu_{eq})/kT. \tag{2.5}$$

Using the Gibbs–Duhem equation these chemical potentials may be related to solute activities or compositions to give

$$\sigma = \ln(a_{ss}/a_{eq}) \tag{2.6}$$

in which a is the activity. For an ideal solution or one in which the activity coefficient is independent of concentration, this reduces to

$$\sigma = \ln(x_{ss}/x_{eq}) \approx (x_{ss} - x_{eq})/x_{eq} \tag{2.7}$$

where the approximation is valid for small values of $(x_{ss} - x_{eq})$.

Chemical engineers, who are predominantly interested in overall yield, often dispense with the use of mole fraction and with the denominator, x_{eq}, and write the supersaturation as

$$\Delta c = c_{ss} - c_{eq} \tag{2.8}$$

with c in some appropriate concentration units. This is usually known as the *concentration driving force*.

When crystallization takes place from the melt, in for example a eutectic system, the supersaturation can be conveniently written in terms of the temperature to give the undercooling ΔT. This can be converted into a thermodynamically precise expression by use of the heat of fusion when

$$\sigma = \frac{\Delta H_f}{R}\left(\frac{\Delta T}{TT_f}\right). \tag{2.9}$$

2.4   Crystallization techniques

The process stage that precedes crystallization is often a chemical reaction. This may be a single-step reaction (e.g. oxidation of paraxylene to yield terephthalic acid or the reaction of ammonia and carbon dioxide to form urea) or it may be the final stage in a complex sequence of chemical reactions as in pharmaceutical or agrochemical production. Whatever the process, the product required generally goes forward to the crystallization either in solution or as an impure liquid above its melting point. This situation has led to the development of two major crystallization techniques.

The first of these relies on crystallization from solutions and is normally referred to as *suspension crystallization*, while the second is concerned with *solidification of melts* and may take the form of either progressive freezing or of solidification of droplets (spray drying or prilling). Both types of process can be operated either continuously or batchwise depending on the product involved. Commodity materials (such as sodium chloride, paraxylene, ammonium nitrate, urea and adipic acid) are produced in high tonnages and production costs are minimized by continuous production in customized plant. Speciality chemicals, on the other hand, are produced in low tonnages using flexible batch processes. The high added value that these materials command in the market means that production costs can be considerably higher than for commodity products.

Suspension processes

Given a solution containing the required product, the mode of crystallization will depend on the yield required and the phase diagram. Referring again to Fig. 2.7, if the starting composition of this solution is x_i then clearly it can be brought into the supersaturated region either by cooling (path a) or by solvent evaporation (path b). In either case, once supersaturated the solution will crystallize and the solution composition will fall to the composition specified by the equilibrium solubility. Further crystallization can be achieved by additional cooling, evaporation, or both.

The slope of the solubility curve largely determines the most suitable method of crystallization. If the solubility varies significantly with temperature, then crystallization by cooling is likely to be a suitable method since a good crystal yield will be obtained by a modest temperature decrease. If, on the other hand, the solubility is relatively insensitive to temperature, as with sodium chloride, even a large decrease in temperature will produce a poor yield and evaporation of solvent would be a preferred technique. In this case the yield will be determined by the solubility of the product at the lowest temperature. In a continuous process this will be unimportant since recycling of liquor streams can take place. For batch processes yields are often maximized by changing the absolute solubility of a product either through the addition of non-solvent (drowning out) or by crystallization of a salt with low solubility.

Suspension crystallization processes are operated to produce particulate materials in appropriate yields and having desired physical characteristics. In subsequent chapters it will become apparent how combinations of equipment design, temperature and supersaturation control, and addition of specific additives, can be used to give specific crystal size distributions, crystal shapes, crystal purities and crystal structures.

Solidification processes

If the product of a chemical reaction is a liquid above its melting point (i.e. a melt) two major techniques are available for conversion to a solid. The largest scale process has been developed in the fertilizer industry for transformation of ammonium nitrate and urea melts into spherical particles (prills) suitable for application to arable land. Molten material is sprayed from the top of a tower in the form of millimetre-sized drops. These cool below their melting point as they fall, so solidifying in flight to give solid particles by the time they reach the base of the tower. Each droplet can be thought of as a micro-crystallizer in

Fig. 2.8 This is a cross-section through an ammonium nitrate prill about 1.5 mm in diameter. It is *polycrystalline*, being an agglomerate of very small crystals. The manufacture of steel, chocolate and candles are all examples of melt crystallization and all these products are polycrystalline.

which the melt transforms to a solid sphere comprising many small crystals as illustrated in Fig. 2.8. No purification can be achieved in this process.

If purification is required then progressive freezing may be used. In this case a cooled surface is placed in contact with the melt and, if the surface temperature is below the melting point, a layer of solid will form at the surface. This layer will grow provided heat is removed at the cooled surface. In a eutectic system a single pure component may be crystallized according to the phase diagram (Fig. 2.5). In a system forming a solid solution (Fig. 2.6) such a solidification process will have to be repeated a number of times in order to move the solid phase composition towards that of the pure component. This is often termed fractional crystallization. In both these cases the product solid is eventually melted off the cooling surfaces to yield a pure liquid which would then be flaked, pastelled or spray cooled if the product were required in solid form.

2.5   Final remarks

In order to define a crystallization strategy we must be able to specify
  (a) the desirable physical characteristics of the product,
  (b) the phase diagram for the crystallizing system, and
  (c) the likely scale of production.
  When particulate characteristics are of highest importance a suspension crystallization process is preferable. If purity is an important aspect then either melt solidification or suspension crystallization may be appropriate. The operation of these processes depends on the generation of supersaturation, the appropriate design of equipment and the economics of the product.

References

Broul M., Nyvlt J. and Sohnel O. (1981) *Solubility in inorganic two-component systems*, Academia, Prague.

Further reading

Findlay A. and Campbell A. N. (1951) *The phase rule and its applications*, 9th edn, Longmans, London.
Mullin J. W. (1992) *Crystallisation*, 3rd edn, Chapter 4, Butterworth-Heinemann, Oxford.
Nyvlt J. (1977) *Solid–liquid phase equilibria*, Elsevier, Amsterdam.
Price G. (1998) *Thermodynamics of chemical processes*, Oxford Chemistry Primers No. 56, Oxford University Press.

3 Nucleation

The process of creating a new solid phase from a supersaturated homogeneous mother phase is called *nucleation* and is central to all types of crystallization.

In this chapter we derive a kinetic expression for the rate of nucleation and then extend the concepts involved to deal with simultaneous nucleation of two phases in a polymorphic system and nucleation in constrained volumes. We also consider heterogeneous systems in which nucleation can be catalysed by the presence of an existing surface in the form of ill-defined 'dust', seeds of the crystallizing material (secondary nucleation) or pre-ordered molecular species such as monolayers or micelles. Finally, techniques for detecting nucleation events are discussed.

3.1 The kinetics of nucleation

The critical size

Let the chemical potential of a molecule in the supersaturated state be μ_{ss} and that in the saturated phase μ_{eq}. In the case of solutions the supersaturation is given (see Section 2.3) by $\ln(x_{ss}/x_{eq})$. In considering the fate of solute molecules in the overall phase change from fluid to solid state it is clear that some molecules end up in the bulk (interior) of the crystal nuclei while others become a part of the surface. In order to quantify the process of phase change consider a cluster or nucleus containing z molecules of which z_b have the properties of the bulk solid and z_s are surface molecules. The free energy of the cluster, g_z, can be written as the sum of the bulk and surface molecular free energies, g_b and g_s:

$$g_z = z_b g_b + z_s g_s \quad \text{or} \quad g_z = (z_b + z_s)g_b + (g_s - g_b)z_s.$$

Introducing the interfacial tension γ between the cluster and the solution and the cluster surface area A we can then write

$$\gamma = (g_s - g_b)z_s/A \tag{3.1}$$

and hence

$$g_z = z g_b + \gamma A. \tag{3.2}$$

Now if the z molecules form a spherical cluster, it follows that

$$A \propto z^{2/3}. \tag{3.3}$$

Writing this in terms of chemical potentials, the free energy of the cluster is

$$g_z = z \mu_b + \beta \gamma z^{2/3} \tag{3.4}$$

in which β is an area shape factor dependent on the nucleus shape and μ_b the chemical potential of a molecule in the bulk of the cluster.

In 1724, Fahrenheit reported that water could be cooled to $-9.4°C$ without freezing. By 1813, Gay-Lussac had shown that the ability of solutions to remain supersaturated without nucleation was very general. It was to be another century before the kinetic formalism given here was developed to explain why it is often so difficult to initiate nucleation in a supersaturated system.

Nucleation, like a chemical reaction, is an activated process with a transition state. However, while in the formation of a covalent bond this transition state may be a bi- or trimolecular complex, in the case of nucleation it is a cluster of a few tens of molecules held together by relatively weak intermolecular forces and packed in a regular way. The size of the cluster and hence the height of the activated barrier depend on the supersaturation.

If the cluster is formed from individual molecules of a substance A present in the bulk fluid phase at mole fraction x_{ss}, then the nucleation event may be written as a quasi equilibrium between monomers and clusters:

$$zA \Leftrightarrow A_z \tag{3.5}$$

and the free energy change per mole of A_z nucleated is $\Delta G = g_z - z\mu$ where μ is the chemical potential of monomers. Now, since $\mu = \mu^o + kT \ln x_{ss}$,

$$\Delta G = (z\mu_b + \beta \gamma z^{2/3}) - z(\mu^o + kT\ln x_{ss}). \tag{3.6}$$

For a saturated solution $x = x_{eq}$ and hence $\mu_b = \mu^o + kT \ln x_{eq}$ so this equation becomes

$$\Delta G = -zkT \ln(x_{ss}/x_{eq}) + \beta \gamma z^{2/3} \tag{3.7}$$

in which $\ln(x_{ss}/x_{eq})$ is the supersaturation as defined in Section 2.3.

Figure 3.1 shows how ΔG depends on z according to Eqn 3.7. The maximum in the curve at the critical size z_c corresponds to the size at which further growth of the cluster leads to a decrease in free energy. For sizes less than this a decrease in free energy can only be achieved by dissolution. Clusters of this critical size are thus called *critical nuclei* and the chance of forming nuclei of this size will depend on the height of the free energy barrier relative to kT. From Fig. 3.1 it is evident that as the supersaturation increases, both the height of the barrier and the value of the critical size decrease. With increasing supersaturation the barrier eventually becomes small enough for nucleation to become spontaneous. The *rate* of nucleation is defined as the rate at which clusters grow through this critical size and so become crystals.

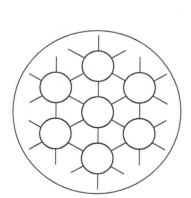

Fig. 3.2 A cluster of molecules during nucleation. Each molecule can form six available intermolecular interactions. In the interior all these are satisfied, while in the surface the molecules are under stress because they cannot realize their full interaction potential.

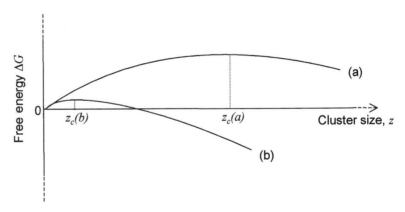

Fig. 3.1 The free energy change as a function of cluster size and supersaturation. Curve (a) at low supersaturation and curve (b) at high supersaturation.

The solubility of small crystals

Because the molecules at the surface of a crystal do not have their intermolecular interactions fully and homogeneously satisfied, they are in a state of tension. We can imagine that a kind of skin is formed that exerts a higher pressure on the molecules in the interior of a crystal than is experienced by those in the fluid phase. If the pressures in the crystal and fluid phases are

p^b and p^f respectively, then this increase in pressure is related to the interfacial tension by:

$$p^b - p^f = 2\gamma/r. \qquad (3.8)$$

As the radius, r, of a crystal decreases, so p^b increases and hence the chemical potential, μ_b, of molecules in the bulk of the crystal increases.

Because at equilibrium the chemical potential of molecules in the crystal and in the solution are equal, it follows that

$$\mu_b = \mu_{eq} = \mu^o + RT \ln x_{eq}(r)$$

and hence that the solubility of the crystals increases with decreasing size. Formally we can write:

$$\mu_b(T, p^b) = \mu_b(T, p^f) + (p^b - p^f)v_c$$

in which v_c is the molar volume in the crystal. Using Eqn 3.8, this becomes

$$\mu_b(T, p^b) = \mu_b(T, p^f) + 2\gamma\, v_c/r.$$

Since at equilibrium $\mu_b = \mu_{eq}$ it follows that

$$\ln x_{eq}(r) = \ln x_{eq}(\infty) + 2\gamma\, v_c/rRT \qquad (3.9)$$

and hence that

$$x_{eq}(r)/x_{eq}(\infty) = \exp(2\gamma\, v_c/rRT). \qquad (3.10)$$

Thus the solubility of a crystal depends on its size, small crystals being more soluble than large. This is referred to as the Ostwald–Freundlich equation, the effect becoming significant as crystal sizes approach the critical size. Figure 3.3 shows the form of Eqn 3.10 for the case of potash alum with the interfacial tension taken as 10 and 20 mJ/m^2.

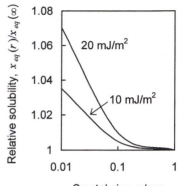

Fig. 3.3 Solubility of potash alum as a function of crystal size at 293 K calculated using Eqn 3.10 using two values of the interfacial tension γ.

This dependence of solubility on size gives rise to *Ostwald Ripening*. A solution that is in equilibrium with large crystals is undersaturated with respect to small ones. Thus in a slurry of crystals having a wide range of sizes it is possible for the small crystals to dissolve and for their mass subsequently to be deposited on the large crystals which grow. In this way, the mean crystal size in the sample increases.

The rate equation

In order to derive an expression for the rate of nucleation we again assume that clusters form by a stepwise aggregation process, which for molecules A with a critical size z_c can be expressed by the overall reaction

$$z_c A \rightarrow A_c.$$

The equilibrium constant for the reaction is

$$K_z = [A]^{z_c}/[A_c].$$

Remembering that equilibrium constants are only disguised free energies we can also write:

$$\ln K_z = -\Delta G_c/RT.$$

Hence the concentration of critical nuclei, $[A_c]$, is related to the activation free energy for nucleation, ΔG_c, by

$$[A_c] = [A]^{z_c} \exp(-\Delta G_c/RT). \qquad (3.11)$$

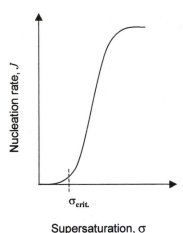

Fig. 3.4 The rate of nucleation as a function of supersaturation. Remember, supersaturation is really free energy. The critical value reflects that fact that creating new surface costs energy.

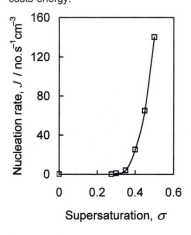

Fig. 3.5 The nucleation rate of sodium perborate from aqueous solutions. Sodium *per*borate is a major source of bleach in washing powders (as in *Per*sil). (After Dugua, 1977.)

Control of crystallinity in polymers is used to improve mechanical properties. Polyester film, for example, is made from amorphous polymer by biaxially stretching it at a temperature below its melting point. This allows controlled crystallization to occur and improves the tear resistance of the film.

By rewriting Eqn 3.2 and assuming spherical nuclei, ΔG_c can be expressed in terms of the crystal radius, r, and interfacial tension, γ:

$$\Delta G_c = (4\pi r_c^3/3)\Delta G_b + 4\pi r_c^2\gamma. \tag{3.12}$$

Since at $r = r_c$, $d\Delta G_c/dr = 0$, it can be shown that the free energy per mole in the bulk crystal, $\Delta G_b = -2\gamma/r_c$. Making use of Eqns 3.11 and 3.12, the concentration of critical-sized nuclei is:

$$[A_c] = [A]^{z_c}\ \exp(-8\pi\ \gamma\ r_c^2/3RT). \tag{3.13}$$

The rate of nucleation J is then this concentration, with the value of r_c given by Eqn 3.10, multiplied by the probability, P, that a critical nucleus will grow to a mature crystal. Thus

$$J = P[A_c] = P[A]^{z_c}\ \exp(-16\gamma^3 v_c^2/3R^3T^3\sigma^2). \tag{3.14a}$$

This may be written in a simplified form that highlights the effect on nucleation rate of the important variables supersaturation, temperature and interfacial tension:

$$J = K_J\ \exp(-B_J\gamma^3/T^3\sigma^2). \tag{3.14b}$$

Figure 3.4 shows the form of this $J(\sigma)$ relationship. At low supersaturations the interfacial tension term dominates and there is insufficient free energy available via the supersaturation to create new surface; hence $J \approx 0$. At some critical value of the supersaturation the nucleation rate increases catastrophically, eventually reaching a maximum value. One of the major triumphs of this result is that it predicts the experimentally known transition from a metastable zone in which, although supersaturation exists, nucleation rates are very low, to the labile zone in which nucleation is spontaneous (see Fig. 2.7).

3.2   Some additional comments

The supersaturation σ_{crit} corresponding to the boundary of the metastable and labile zones can be defined if the critical nucleation rate is chosen to be some specific value, say one nucleus per second per unit volume of fluid. This is done by taking the logarithmic form of the nucleation equation when

$$\sigma_{crit} = (16\gamma^3 v_c^2/3R^3T^3)^{0.5}. \tag{3.15}$$

σ_{crit} thus represents the width of the metastable zone; within this zone crystals can be grown without additional nucleation occurring. This principle is important in seeded industrial crystallizers and for the growth of large single crystals for electronic devices.

Generally, primary nucleation is important in the production of speciality chemicals such as dyes, pharmaceuticals, photographic chemicals, pigments and catalysts where the solute solubilities tend to be low and so the supersaturations are high. When we consider solidification from melts of polymers, inorganic or organic materials, cooling below the melting point (or undercooling as it is often referred to) increases the supersaturation. But this cooling also reduces the available thermal energy so that it is possible for transport processes such as diffusion and viscous flow to become rate controlling. Thus, polymers can often be supercooled to form amorphous solids that only crystallize on subsequent heating.

In order to account theoretically for this *transport resistance* to nucleation, Eqn 3.14 can be modified to include an activation free energy for the transport resistance, ΔG_{tr}:

$$J = P[A_c] \exp\{(-16\pi\gamma^3 v_c^2/3R^3T^3\sigma^2) + \Delta G_{tr}/RT\}. \qquad (3.16)$$

In this situation the nucleation rate passes through a maximum with increasing supersaturation as shown in Fig. 3.6 for the crystallization of the biopolymer polyhydroxybutyrate.

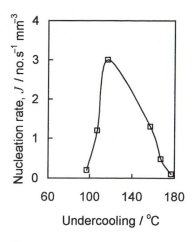

Fig. 3.6 The nucleation rate of polyhydroxybutyrate as a function of undercooling, $T_f - T$. (After Barham, 1992.)

3.3 Nucleation in polymorphic systems

For crystallizing systems where more than one phase can crystallize (polymorphs or solvates for example), Ostwald, in 1897, proposed a generalized rule based on observations of many systems. Sometimes referred to as the *Rule of Stages*, this postulated that a crystallizing system progresses from the supersaturated state to equilibrium in stages, each stage representing the smallest possible change in free energy. Thus, a polymorphic system would move through each possible polymorphic structure before crystals of the most stable phase appear. In the simple dimorphic system of Fig. 3.7 this means the initial appearance of crystals of phase I, followed by their transformation to form II.

The justification for Ostwald's rule can be explored using Fig. 3.7 and Eqn 2.7. A solution of composition x_B at temperature T_i is supersaturated only with respect to the stable phase II, whilst from a solution of composition x_i both phases I and II can precipitate. We will examine the latter case. Two supersaturations can be defined, both with respect to the more stable phase II. The first is the initial supersaturation:

$$\sigma_i = (x_i - x_{II})/x_{II}$$

while the second is the supersaturation in a solution saturated with phase I:

$$\sigma_x = (x_I - x_{II})/x_{II}.$$

Nucleation rate expressions can be written for each phase:

$$J_I = K_{J,I} \exp\left[-B_I/(\sigma_i - \sigma_x)^2\right] \qquad (3.17a)$$

and

$$J_{II} = K_{J,II} \exp\left[-B_{II}/\sigma_i^2\right] \qquad (3.17b)$$

with the two values of B being given by the appropriate values of the term $16\pi\gamma^3 v^2/3R^3T^3$. Both $K_{J,I}$ and $K_{J,II}$ are functions of x_{eq} and T. By defining the dimensionless variables:

$$a = \sigma_x/B_{II}^{1/2}, \quad b = (B_I/B_{II}) \quad \text{and} \quad c = [a/\ln(K_{J,II}/K_{J,I})]^{1/3},$$

Remember that if one molecular structure gives rise to more than one crystal structure the material is termed *polymorphic*. Note also that the terms phase, form and polymorph are often used interchangeably.

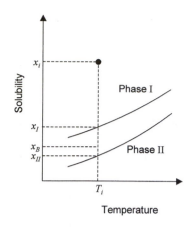

Fig. 3.7 The solubility curves of two polymorphic forms I and II.

Eqns 3.17a and 3.17b can be solved simultaneously to explore the possible nucleation behaviour in such a system. It can be shown (Cardew and Davey, 1993) that:

1. If $K_{J,I} > K_{J,II}$, then above some value of supersaturation, the metastable phase I has the highest nucleation rate whereas below this value phase II appears more rapidly.
2. If $K_{J,II} > K_{J,I}$ and $(1 - a/c)^3 < b$, the stable phase II has the higher nucleation rate at all supersaturations.
3. If $K_{J,II} > K_{J,I}$ and $(1 - a/c)^3 > b$, the metastable phase has the higher nucleation rate only over the intermediate range of supersaturations.

All three cases are shown graphically in Fig. 3.8. Ostwald's rule implies that $J_I \gg J_{II}$. It is clear that this condition is not generally valid. The phase having the higher nucleation rate can vary with supersaturation and depends on both surface parameters (which are included in the values of B) and bulk parameters such as x_{eq} and T. A consequence of this analysis is the need to exercise great caution when processes are being developed to crystallize a specific polymorph.

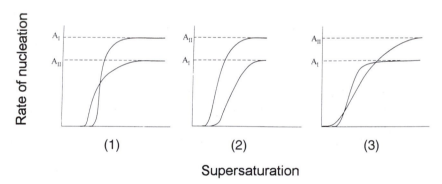

Fig. 3.8 Nucleation kinetics in a system with two polymorphs, I and II, showing the three possible cases.

3.4 Heterogeneous nucleation

Further modification of the basic nucleation equation is also possible to account for primary nucleation in 'dirty' systems. The presence of foreign bodies or 'catalytic' surfaces can induce nucleation at supersaturations lower than those required for homogenous nucleation. A surface allows adsorption of crystallizing material and lowers the value of ΔG_c, the extent of this reduction depending on the degree to which the catalysing body mimics the structure of the crystallizing material. This type of nucleation is referred to as *heterogeneous*.

At one extreme, this process occurs by non-oriented adsorption of solute onto any existing surface until it resembles a crystal; at the other, specific structural relationships exist between catalyst and crystallizing solute in processes which are *epitaxial*. In this latter case the surface structure of an added particle induces oriented adsorption of solute molecules so as to mimic its crystal structure. Figure 3.9 shows on example of this for the nucleation of polyhydroxybutyrate in the presence of crystalline particles of saccharin. The enhancement of the nucleation rate in this case is evident by comparing these data with those in Fig. 3.6.

One further example of epitaxially catalysed nucleation has recently been studied in detail. A monolayer of an amphiphilic molecule is spread over the surface of a supersaturated solution and crystallization allowed to proceed. For the crystallization of ice, calcite, barium sulphate and glycine it has proved possible to induce oriented nucleation at the interface between such a monolayer and the supersaturated subphase. In the case of ice the nucleation temperature can be raised from $-10°C$ to $-0.1°C$ by use of a monolayer of an *n*-alkanol. This catalytic effect arises from the structural match between that of the ice lattice and the packing of hydroxyl groups in the alcohol monolayer (Weisbuch *et al.*, 1994).

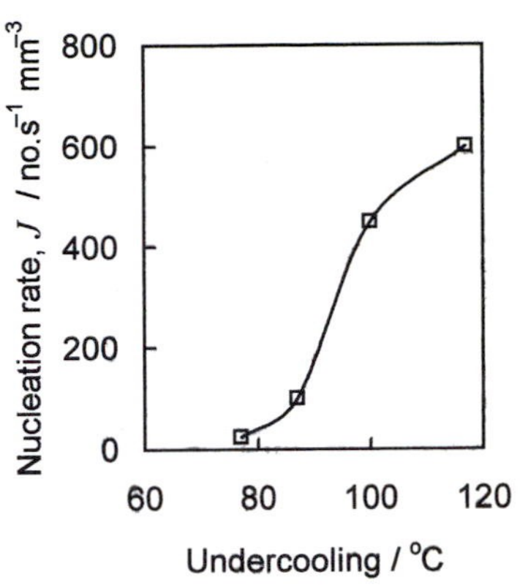

Fig. 3.9 The nucleation kinetics of polyhydroxybutyrate in the presence of saccharin crystals. (After Barham, 1992.) Compare this with Fig. 3.6.

3.5 Secondary nucleation

The best match between substrate and crystallizing solute exists when the substrate is in fact a seed crystal of the solute. This situation occurs when batch crystallizers are 'seeded' and is always present in a continuous crystallizer (see Chapter 8). Nucleation of new crystals induced only because of the prior presence of crystals of the material being crystallized is termed *secondary nucleation*. This nucleation mechanism generally occurs at much lower supersaturations (or undercoolings) than primary homogeneous or even heterogeneous nucleation.

Secondary nucleation can occur by a number of different mechanisms. If dry crystal seeds are placed in a supersaturated solution they will often shed particles of crystalline dust that had been adhering to their surfaces. These then become new centres for growth, or 'secondary nuclei'. This is detrimental to the effective seeding of batch crystallizers that will be described in Section 8.5. Crystals that grow in the form of needles or dendrites are often rather fragile and small parts of the crystal may break, again forming secondary nuclei. There is also evidence that the shear forces imposed on a crystal face by the solution flowing past it can be sufficient to produce secondary nuclei from the surface—so-called *shear nucleation*.

The most important and commonly encountered mechanism of secondary nucleation is *contact nucleation*, sometimes also referred to as collision nucleation or collision breeding. Contacts between a growing crystal and walls of the container, a stirrer or pump impeller, or other crystals result in contact nucleation. It is now recognized that for materials of high or moderate solubility this is the most significant nucleation mechanism in crystallizers.

Many experiments have shown that the nucleation rate in crystallizers depends not only on the supersaturation but also on the concentration of crystals in suspension, M_T, and on some measure of the hydrodynamic interactions between the crystals and the solution, for example stirrer speed, N,

Whereas carefully purified water can be undercooled to perhaps $-30°C$ before primary homogeneous nucleation occurs, and tap water, nucleating by a heterogeneous mechanism, can be cooled to perhaps $-9°C$, a continuous crystallizer containing a retained bed of ice crystals will operate perfectly well at -2 or $-3°C$, secondary nucleation producing the necessary number of new nuclei.

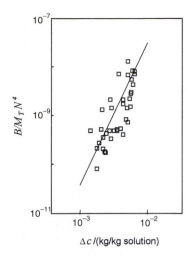

Fig. 3.10 Secondary nucleation kinetics for potash alum in a 1.3 dm³ crystallizer. The data are plotted in a form suggested by Eqn 3.18. In this case $j = 1.0$, $k = 4.0$ and $b = 2.52$. (After Garside and Jancic, 1979.)

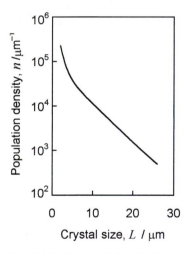

Fig. 3.11 The initial size distribution of secondary nuclei of potash alum. A Coulter Counter® was used to measure the size with 2.2 μm being the lower limit. (After Garside *et al.*, 1979.)

or average power input to the solution, ε. The effect of these variables can be expressed empirically as a power law function:

$$B = k_b M_T^j N^k \Delta c^b \tag{3.18}$$

where the symbol B is used to represent the *secondary* nucleation rate and N (a stirrer, or pump impeller, rotation rate) is taken as a measure of the fluid mechanics interactions and of the power input.

Typical values of b lie between about 1 and 2.5. These are much lower than the values that would be found if the primary nucleation rate equation (Eqn 3.14) were fitted to a power law function of Δc. The influence of crystal concentration points directly to the importance of crystal collisions. Most values of j are close to unity, suggesting the dominance of collisions between crystals and the vessel walls or, more probably, between crystals and the stirrer rather than between two crystals. Specific values of k predicted by various semi-empirical models of contact nucleation in crystallizers are in the range 2 to 4, as are the majority of experimentally determined values. An example of experimental data is shown in Fig. 3.10.

Much work has been directed at clarifying the mechanisms by which secondary contact nuclei are produced. A series of elegant experiments (Denk and Botsaris, 1972) made use of the left- and right-handed morphological properties of sodium chlorate to distinguish between nuclei originating from a parent crystal and from the solution. Under a wide range of conditions it was shown that nuclei are formed from the crystal surface rather than the solution. Direct observation of micro-attrition at a crystal surface demonstrated that large numbers of fragments were produced directly into sizes between 1 and 10 μm with some larger fragments up to 50 μm (Garside *et al.*, 1979). An example of the size distribution of secondary nuclei produced by contact nucleation is shown in Fig. 3.11.

The great interest in secondary nucleation arises from its importance in industrial crystallizers. In this context the challenge is to develop a mechanistic description of secondary nucleation to allow prediction of nucleation rates in commercial scale plants from experiments on a smaller scale. Most such attempts start from the view that the overall secondary nucleation rate is the product of three functions. The first describes the rate at which energy is transferred to a crystal by an appropriate collision mechanism. The second function represents the number, or perhaps the size distribution, of fragments produced per unit of transferred energy, while the third function is a survival term representing the fraction of these particles that survive to become nuclei and subsequently grow to populate a size distribution. All these approaches give rise to equations having the general form of Eqn 3.18 in which the value of k_b is a specified function of the vessel geometry while the values of the exponents j and k are determined by the particular assumptions of the model.

3.6   Nucleation in constrained volumes

Although the kinetic rate equation (Eqn 3.14) applies to a situation in which the supersaturated phase is free from any contaminating catalytic surfaces, we have seen how nucleation in real systems is often of a heterogeneous nature. In order to study homogeneous nucleation experimentally, Turnbull (1950) suggested that if a bulk sample of liquid is broken down into a large number of isolated drops, perhaps 10–100 μm in size, then the foreign particles responsible for heterogeneous nucleation would be localized in only a small number of these drops. The remaining drops may then be used for studying homogeneous nucleation (Fig. 3.12).

The original studies exploring this idea focused on the solidification of metals but the technique has since been extended to the study of ice formation, solidification of molecular materials and crystallization from solutions. Typically the material to be studied is emulsified as a melt or solution in a continuous phase using an appropriate surfactant. The onset of nucleation is recorded in some appropriate way (see Section 3.7). Fig. 3.13 shows the crystallization curves for solutions of glyceryl tristearate in paraffin oil which were emulsified in water using a range of surfactants. It is apparent that the emulsified systems will support higher undercoolings than the non-emulsified system. It has also been shown that the choice of emulsifier has some effect on the nucleation process.

The use of small volumes as a means of stabilizing supersaturated liquids is of practical importance in the formulation of foods, creams and explosives, and in biological systems where freezing of water in particular is often prevented in this way.

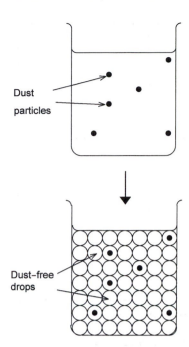

Fig. 3.12 A beaker of solution contains lots of submicrometre dust particles. If the solution is transformed into many small drops some of them will be dust free.

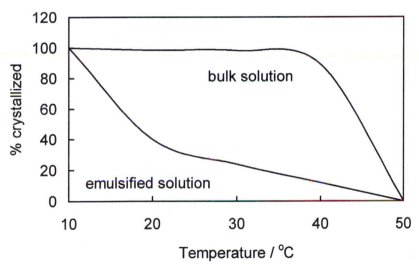

Fig. 3.13 Comparison of the nucleation of glyceryl tristearate in emulsified systems with nucleation in the bulk phase. (After Skoda and van den Tempel, 1963.)

3.7 Detecting nucleation events

How do we know if a system has nucleated? The usual approach is to measure continuously an appropriate property of the system as nucleation proceeds. A number of properties can be used:

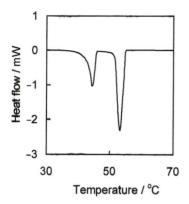

Fig. 3.14 A differential scanning calorimetry (DSC) trace showing the crystallization of stearyl alcohol at 55°C and its subsequent polymorphic transformation at 44°C. DSC measures the heat flows into and out of a sample as its temperature is raised or lowered in tandem with a reference sample. The corresponding exo- and endotherms relate to crystallization and melting.

Temperature Since nucleation reduces the free energy of a system it is usual for heat to be evolved. This may be detected either by direct measurement of the temperature change or by differential scanning calorimetry (Fig. 3.14).

Volume Solid phases have different densities from their mother phases and so it is possible to use dilatometry to measure the volume of a crystallizing system with time. The data given for glyceryl tristearate emulsions in Fig. 3.13 were measured using this technique.

Optical transmittance Nucleation produces a dispersion of fine crystals. Consequently the optical properties of the system undergo a step change at the point of nucleation. This allows the use of turbidometry (light scattering) for the detection of nucleation.

Concentration measurement In a system containing two or more components the nucleation of the solute is accompanied by a decrease in the solution concentration. This can be monitored by following the change in a concentration-dependent physical property such as conductivity, density or refractive index.

In isothermal systems, these techniques can be used to estimate the time lapse between the initiation of supersaturation and the onset of nucleation. This is referred to as the *induction time*, τ_{ind}, and is often used as a measure of nucleation rates through the relation

$$J \propto 1/\tau_{ind}. \tag{3.19}$$

Combining this with Eqn 3.14 it is evident that

$$\ln(1/\tau_{ind}) \propto \gamma^3/T^3\sigma^2. \tag{3.20}$$

Induction times measured as a function of supersaturation can thus be used to estimate the important parameter γ, the crystal–liquid interfacial tension. Figure 3.15 shows typical data measured for sucrose crystallization which yield a value of 4.7 mJ/m² for the interfacial tension.

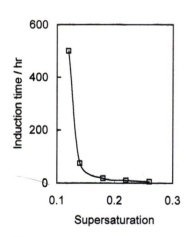

Fig. 3.15 Induction times as a function of supersaturation for the nucleation of sucrose from aqueous solution. (After Dunning and Shipman, 1954.)

References

Barham P. J. (1992) *J. Mater. Sci.* **19** 3826.

Cardew P. T. and Davey R. J. (1993) *Faraday Soc. Discuss* **No. 95** 160.

Denk E. G. and Botsaris G. D. (1972) *J. Cryst. Growth* **13/14** 493.

Dugua J. (1977) Influence des agents tensio-actifs sur la cristallisation du perborate de sodium. Thesis, Université *d'Aix* Marseille.

Dunning W. J. and Shipman A. J. (1954) Nucleation in sucrose solutions. *Proc. Agric. Industr. 10th Int. Conf.*, Madrid, Spain, p. 1448.

Garside J. and Jancic S. J. (1979) *AIChE J.* **25** 948.

Garside J., Rusli I. T. and Larson M. A. (1979) *AIChE J.* **25** 57.

Grainger L. (1999) Crystallization of organic compounds in cyclomethicone. MPhil. Thesis, UMIST.

Skoda W. and van den Tempel M. (1963) *J. Coll. Sci.* **18** 568.

Turnbull D. (1950) J. Chem. Phys **18** 198.

Weisbuch I., Popovitz-Biro R., Leiserowitz L. and Lahav M. (1994) in *The lock and key principle*, J. P. Behr (ed.), p. 173, Wiley, New York.

Further reading

Mullin J. W. (1992) *Crystallization*, 3rd edn, Chapters 5 and 6, Butterworth-Heinemann, Oxford.

Zettlemoyer A. C. (ed.) (1969) *Nucleation,* Marcell Dekker, New York. Chapters 1 and 5 are particularly relevant.

4 Crystal growth

4.1 Crystal surfaces

When a crystal surface is exposed to a supersaturated environment, the flux of growth units (atoms, ions, molecules) to the surface exceeds the equilibrium flux so that the number of growth units joining the surface is greater than that leaving. This results in growth of the surface. The ability of a surface to capture arriving growth units and integrate them into the crystal lattice is, among other things, dependent upon the strength and number of interactions that can form between the surface and the growth unit.

The importance of solid state structural effects is illustrated for a simple two-dimensional crystal in Fig. 4.1. In the interior of this crystal each molecule is bound to four of its nearest neighbours as indicated by the solid lines. Two surfaces bound this crystal, one having two intermolecular interactions available at the surface (1) and one with only a single interaction (2). When molecules join surface 1 the system gains twice as much energy as when they join 2. For this reason surface 1 will grow faster than surface 2.

In three dimensions (Fig. 4.2) it is clear that a growth unit may form a maximum of three bonds with a crystal surface. Using the classification first introduced by Hartman and Perdock the following definitions are made: a surface with which three bonds are possible is a *kinked* or K face, a surface with which two bonds are possible is a *stepped* or S face, and a face with which just one surface bond is possible is a *flat* or F face. If we assume that the linear growth rate, v, of a face is proportional to the total binding energy of a growth unit to that surface, it would be expected that

$$v_K > v_S > v_F.$$

Thus, as indicated by the dotted outline in Fig. 4.1 the final shape (also termed the morphology, form or habit) of a crystal, a subject discussed in more detail in Chapter 5, will be defined by the slowest growing F faces.

Crystal growth theory, therefore, is concerned with the mechanisms by which F faces grow. We also need to recognize that F faces are characterized not only by the fact that they offer a single binding interaction to incoming growth units but also by the way in which other structural bonds lie parallel to the plane of the face. Given that the strength of even the strongest intermolecular interaction (a hydrogen bond) is only about 20 kJ/mol, compared to perhaps 200 kJ/mol for the strength of a covalent bond, the direct addition of molecules to flat surfaces would never proceed quickly enough to explain why crystals grow at measurable rates.

The growth of such faces can in fact only proceed at a reasonable rate if some means of creating growth sites with higher interaction energies can be envisaged. This might involve, for example, the creation of step and kink sites

The nature of a growth unit remains the subject of some debate: in some cases it will be a single molecule, in others some pre-association may have occurred in solution and the growth unit may be a dimer or an even larger aggregate.

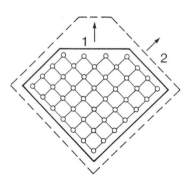

Fig. 4.1 A two-dimensional crystal. The dotted outline indicates the crystal shape after a period of growth. Surface 1 grows faster than surface 2. As a result surface 1 gets smaller and surface 2 is enlarged. This leads to the general conclusion that the importance of a face in the overall crystal morphology varies as the inverse of its growth rate—a crystal is eventually bounded by the slow-growing faces.

on an F face. Since the energetics of this process will depend not only on the solid state structure but also on the supersaturated fluid adjoining it, we need to have some description of solid–fluid interfaces.

4.2 Crystal–fluid interfaces

The most useful model of the crystal–fluid interface is the so-called 'multilayer' model attributed to Temkin (1966). Solid and fluid are divided into blocks of equal size (in a melt each block could be identified with a molecule) and the interfacial region may consist of any number of layers. Figure 4.3 shows a typical interface depicted in this way. The model can easily be used to deduce the energy change, ΔE, occurring when a perfectly flat surface is roughened by removing one block from the surface and using it to start a new layer. This is an important quantity for crystal growth since if this is an energetically favourable process it would be a means by which sites with higher interaction energies could be formed. To explore this concept further three energies are defined:

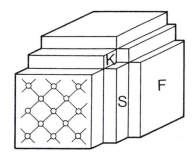

Fig. 4.2 A three-dimensional crystal showing K, S and F faces.

ϕ_{ss}, the interaction energy between the sides of each solid block,
ϕ_{ff}, the interaction energy between the sides of each fluid block, and
ϕ_{sf}, the interaction energy between the sides of fluid and solid blocks.

In the process depicted in Fig. 4.3 a bookkeeping calculation of interactions broken versus those made allows ΔE to be calculated as:

$$\Delta E = 2\phi_{ss} + 2\phi_{ff} - 4\phi_{sf}. \tag{4.1}$$

Redefining this in dimensionless terms as:

$$\alpha = \Delta E / kT \tag{4.2}$$

yields a parameter, the α-factor, which reflects the ease with which a surface can form sites with multiple binding interactions and hence is an indication of the ease with which a surface can grow. If the value of the α-factor is low then growth can proceed easily with many growth sites always present. As the

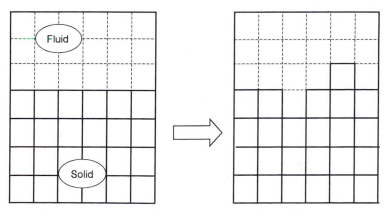

Fig. 4.3 Roughening a solid–fluid interface.

The rate of growth of a crystal can be defined by the rate at which either its mass or one of its dimensions changes with time. In order to differentiate the behaviour of different faces of a crystal it is common to measure the linear rate of advance normal to the plane of a specific face. A typical value would be 0.5 mm/hr ($= 1.4 \times 10^{-7}$ μm/s), equivalent to about 10^{20} molecules per hour per mm^2 of crystal surface. We give this linear face growth rate the symbol v.

value increases growth becomes more difficult and special mechanisms have to be envisaged by which necessary growth sites can be created.

The calculation of α from measurable quantities may be achieved by use of equations that are now in common use:

$$\alpha = \xi \frac{\Delta H}{RT} \tag{4.3}$$

for vapour and melt growth with ΔH the heat of sublimation or fusion, and

$$\alpha = \xi\big[(\Delta H_f/RT) - \ln x_{eq}\big] \tag{4.4}$$

for solution growth with ΔH_f being the heat of fusion and x_{eq} the solubility. ξ is a crystallographic factor that describes the intermolecular interactions in the crystal surface of interest. Thus:

$$\xi = E_{sl}/E_{ss} \approx z_s/z_t \tag{4.5}$$

in which E_{sl} is the total interaction energy per molecule in the layer of the growth face and E_{ss} is the total crystallization (or lattice) energy. z_s and z_t are the respective numbers of nearest neighbours. Values of α typically range from 2 up to 20.

In three dimensions, the block-structured interface appears as in Fig. 4.4 with ledge sites (having an interaction energy of ϕ_{ss}), step sites (with interaction energy $2\phi_{ss}$) and kink sites (of interaction energy $3\phi_{ss}$) indicated. These kink site positions offer enhanced binding to a growth unit and hence are crucial to the growth process. In a real crystal such sites may be identified with specific intermolecular interactions. Figure 4.5 is a two-dimensional projection of the crystal structure of the dicarboxylic acid, adipic acid, $COOH(CH_2)_6COOH$. This is an important intermediate in the production of nylon and it is purified by crystallization. Note how molecules join the (100) face by formation of intermolecular hydrogen bonds and how the same interaction is utilized by molecules joining a step site on the (001) face.

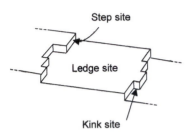

Fig. 4.4 A three-dimensional crystal surface, showing a ledge site, a step site and a kink site.

If a crystal surface is considered to be created by cleaving through the crystal structure then it is apparent that any number of surfaces may be created, each exposing different parts of the constituent molecules to the growth environment. In order to relate a surface to the structure of the crystal a system of *Miller Indices* is used. These define the orientation of the surface in relation to the crystallographic unit cell. Each face is designated with three numbers, these being the inverse of the intersections of that face with the three crystallographic axes, a, b and c. Thus, a surface (110) is the surface that cuts the a- and b-axes one unit cell length from the origin but which is parallel to the c-axis. Different types of brackets have different meanings: {hkl} refers to a set of symmetry-related faces all having the same surface structure; (hkl) refers to a specific single face; and [hkl] is a direction perpendicular to the (hkl) face.

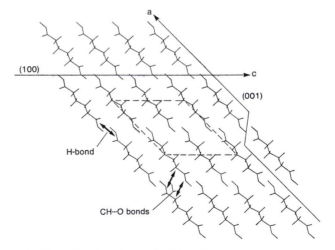

Fig. 4.5 The (100) and (001) surfaces of adipic acid.

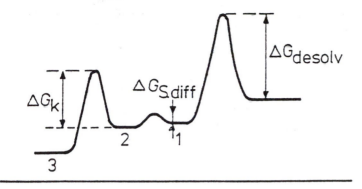

Fig. 4.6 Free energy barriers to be overcome during crystal growth.

As a crystal surface grows, growth units are transferred from the fluid to the solid state. Figure 4.6 illustrates schematically the free energy barriers to be overcome during the growth process. Thus the growth unit, initially solvated, loses some of its solvent molecules upon passage over the first barrier, ΔG_{desolv}. This partially desolvated growth unit then enters the adsorption layer (position 1) from where it may diffuse across the surface (ΔG_{sdiff}) to a step (position 2) and undergo further desolvation to enter a kink site at position 3 (ΔG_k).

4.3 Growth mechanisms

We now consider in more detail the possible pathways by which a molecule passes from the solution to become integrated into a lattice position on a growing face.

Continuous growth

If the value of α is less than about 3, the energy required to form a step is low and so the surface of the crystal will contain many kink and step sites. This suggests that every growth unit arriving at the surface will find a growth site. In this case, the linear growth rate normal to the surface, v, takes on a very simple form, namely:

$$v = k_{CG}\sigma \qquad (4.6)$$

over the entire range of supersaturation.

Surface nucleation

As α increases ($3 < \alpha < 5$) the inherent roughness of the interface decreases and some of the growth units that arrive at the surface do not find a growth site. These either return to the fluid phase or join other adsorbed growth units to form surface islands or nuclei. As Fig. 4.7 shows, the circumference of such islands becomes the source of new step and kink sites at which additional units can join the surface. In this way the islands spread laterally across the surface and growth occurs normal to the plane of the face. The expression for the linear growth rate now takes the form:

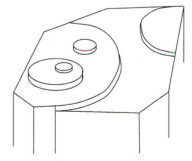

Fig. 4.7 The creation of nuclei on a surface during crystal growth.

$$v = f_1(\text{number of critical size nuclei formed per unit time}) \times f_2(\text{step height})$$
$$\times f_3(\text{step velocity}).$$

This results in the kinetic expression

$$v = k_{SN}\sigma^{5/6} \exp\left\{\left(-\frac{\pi}{3\sigma}\right)\left(\frac{\gamma_e}{kT}\right)^2\right\} \qquad (4.7a)$$

which can be written in the simplified form

$$v = k_{SN}\sigma^{5/6} \exp\left(-\frac{B}{\sigma}\right). \qquad (4.7b)$$

The exponential term in Eqn 4.7 is similar to that in the homogeneous primary nucleation equation (3.14). It arises from the need to create a critically sized two-dimensional nuclei with edge tension γ_e.

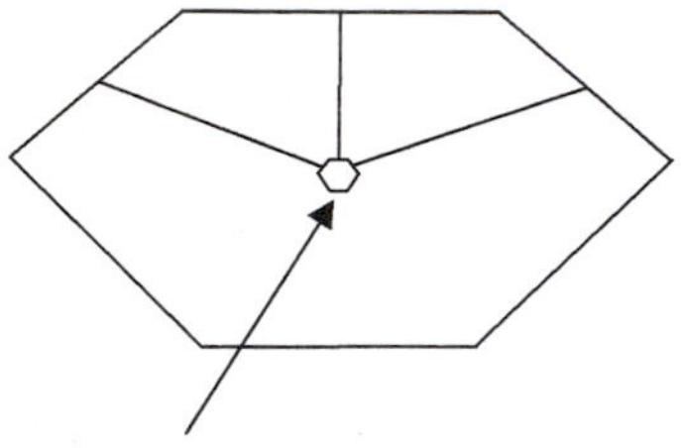

Spiral growth

As α rises above 5, the enhanced intermolecular interactions in the plane of the interface result in a molecular surface that is flat. This is reflected in an increasing value of γ_e which begins to inhibit surface nucleation, particularly at low supersaturation. Growth can now only take place if a step can be created by some energetically 'cheap' process. Built-in lattice defects, in particular screw dislocations (see Fig. 4.8), provide such a route.

A dislocation is the result of the stresses that occur during crystal growth, particularly when growth takes place onto seed crystals, around heteronuclei and around liquid inclusions. Usually one part of the crystal lattice becomes misaligned with respect to the rest of the crystal.

Screw dislocations emerge on the growing crystal faces and the points of emergence are characterized by steps at which growth can then take place. This possibility was first postulated by Frank (1949) and represented a major advance in understanding the mechanism of crystal growth. For crystal growth the important feature of the emergent step is that it extends over only a part of the surface and during growth winds up into a spiral to create a growth hillock (Fig. 4.9).

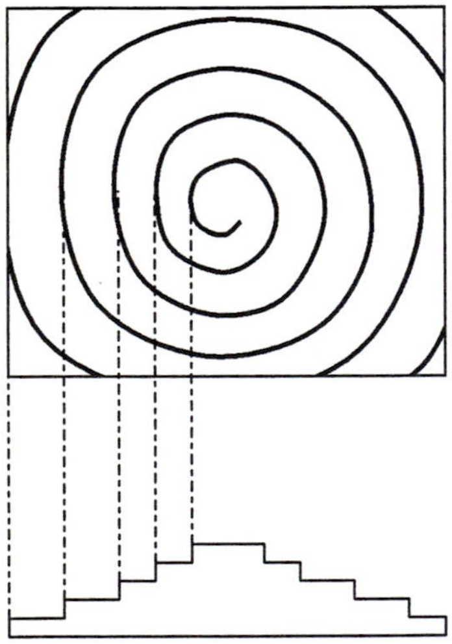

The theoretical expression for the growth rate of a surface growing in this manner is again rather complex although simple in concept. It has the form

$$v = f_1(\text{step velocity}) \times f_2(\text{step height}) \times f_3(\text{step density}).$$

The step velocity is related to the flux of growth units that enter kink sites. Burton, Cabrera and Frank (BCF, 1951) postulated that this may be controlled either by diffusion from the bulk of the solution directly into kink sites, or by two-dimensional diffusion across the crystal surface. The step density is related to their spiral nature. As the supersaturation increases the spiral winds itself tighter, thereby increasing the step density.

The final BCF expression relating growth rate and supersaturation is:

$$v = k_{SG}\frac{\sigma^2}{\sigma_1}\tanh\left(\frac{\sigma_1}{\sigma}\right) \qquad (4.8)$$

Nucleus

Fig. 4.8 A typical dislocation structure as revealed by X-ray analysis is shown here. The dark lines indicate the position of dislocations. Note how these faults terminate at crystal surfaces.

F. C. Frank was a physicist working at Bristol University. He had played an important role as a scientific advisor to the government during World War II and his seminal work on crystal growth was supported in part by the chemical company ICI.

Fig. 4.9 A spiral growth hillock viewed from the side and from above.

in which $k_{SG} \propto \exp(-\Delta G_{desolv}/kT)$ and $\sigma_1 \propto \gamma_e/s$ where s is the strength of the dislocation source. Equation (4.8) has the features that when $\sigma \ll \sigma_1$ the relationship between growth rate and supersaturation is parabolic, while for $\sigma \gg \sigma_1$ it is linear.

An important consequence of spiral growth is that each crystal can have its own unique growth rate determined by its specific dislocation structure.

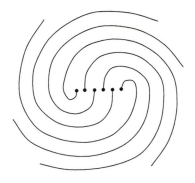

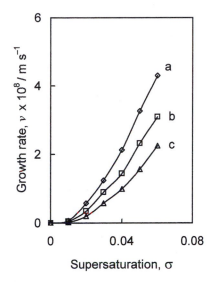

Fig. 4.10 It is possible for separate dislocation sources to interact with one another. In the case shown here six dislocations are interacting. The effect of increasing the step density by the factor *s*, the strength of the dislocation source (six in the case), is to enable the surface to grow faster.

Fig. 4.11 Growth rate data for three different crystals, a, b and c, of ammonium dihydrogen phosphate are shown here. Each crystal grows at its own rate with a spiral growth mechanism. Such kinetic individuality is an obvious consequence of a structure-sensitive growth mechanism (Davey *et al.*, 1979).

Identification of growth mechanisms

The form of the theoretical relationships, $v(\sigma)$, between growth rate and supersaturation predicted by these three mechanisms differ significantly. Figure 4.12 compares the three equations. Experimental kinetic measurements can be used to identify specific growth mechanisms.

Two broad classes of experiment have been carried out to explore and test the applicability of these kinetic models. In the first a large single crystal is mounted in a supersaturated environment and the position of an advancing face monitored with time. From this, the face growth rate can be determined. By repeating this experiment over a range of supersaturations, it is possible to obtain data that can be used to test the possible kinetic models.

Figure 4.13 shows data measured on the $\{0\overline{1}\overline{1}\}$ and $\{011\}$ faces of α-resorcinol (Davey *et al.*, 1988) growing in aqueous solution at 20°C. The $\{0\overline{1}\overline{1}\}$ faces exhibit a BCF-like curve (Eqn 4.8) with $k_{SG} = 10.7$ μm/s and $\sigma_1 = 0.097$, while the $\{011\}$ faces grows by a surface nucleation process (Eqn 4.7b) with $k_{SN} = 130$ μm/s and $B = 0.36$.

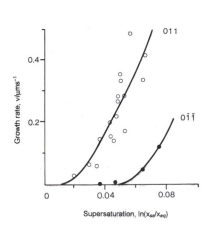

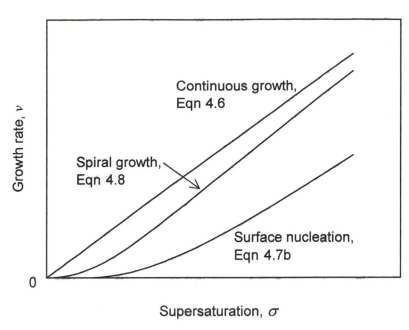

Fig. 4.13 The growth kinetics of α-resorcinol.

Fig. 4.12 Growth rate versus supersaturation curves for three different growth mechanisms.

Various optical microscopic techniques such as interference contrast, phase contrast, interferometry and reflection microscopy have been used for *in situ* studies of growing crystal surfaces. From such observations the existence of growth spirals has been confirmed and the general form of the BCF equations substantiated. More recently, atomic force microscopy (AFM) has been used. Fig 4.14 shows an AFM image of a spiral on the surface of a protein crystal; the step is only 3.4 nm high.

4.4 The importance of mass transport

Before growth units can become incorporated into the crystal lattice by one of the mechanisms described above they must be transported from the bulk of the solution to the crystal–solution interface. The solute concentration at the interface, c_i, is thus less than that in the bulk solution, c_∞, as a result of the diffusive or convective mass transfer resistance.

Many sets of experimental data demonstrate the role of mass transfer in crystal growth from solution. Figure 4.15 illustrates one example; increasing the solution velocity past the crystal results in an increase in the growth rate of the {111} faces of a potash alum crystal because the mass transport resistance is reduced at the higher solution flow rate.

Mass transport and surface integration take place consecutively (i.e. in series with one another) and therefore at steady state the solution concentration at the crystal surface adjusts to a value such that the rates of the two steps are equal.

Fig. 4.14 Spiral on protein crystal–an AFM image (Donohoe *et al.*, 1999). The image 15 × 15 μm.

For most of the growth models discussed above the growth rate can be represented with good accuracy by the power law expression

$$R_G = k_r(c_i - c_{eq})^r \qquad (4.9)$$

in which k_r is the growth rate coefficient and R_G, the overall mass growth rate, is expressed as the total mass flux to the crystal surface and can be related to the face growth rate v by

$$R_G = \rho_{cry}v. \qquad (4.10)$$

It is evident from Eqns 4.6, 4.7 and 4.8 together with Fig. 4.12 that r is usually between 1 and 2.

The diffusion step will be represented by

$$R_G = k_d(c_\infty - c_i) \qquad (4.11)$$

where k_d is the mass transfer coefficient. Elimination of the unknown interfacial concentration c_i between these two growth rate equations gives

$$R_G = k_r\big[(c_\infty - c_{eq}) - R_G/k_d\big]^r. \qquad (4.12)$$

If the surface integration step is first order (i.e. $r = 1$), Eqn 4.12 becomes

$$R_G = k_G(c_\infty - c_{eq}) \qquad (4.13)$$

where

$$\frac{1}{k_G} = \frac{1}{k_d} + \frac{1}{k_r}. \qquad (4.14)$$

This is the well-known expression for the addition of two resistances in series, the coefficients representing conductances and their reciprocals the resistances. The mass transfer and surface integration steps can thus be thought of as two processes taking place in series with k_G being the overall growth rate coefficient.

The concept of effectiveness factor, well established in reaction engineering (e.g. Levenspiel, 1972), is useful in characterizing the relative importance of the diffusional and integration resistances (Garside, 1991). The extent to which the diffusional resistance influences the growth rate can be represented in terms of the surface integration effectiveness factor, η_i, defined as

$$\eta_i = \frac{\text{growth rate at the interface conditions (i.e. the actual growth rate)}}{\text{growth rate if the interface were exposed to the bulk conditions}}.$$

The maximum value of η_i is unity, which occurs when the growth is totally controlled by the integration step.

It can be shown from Eqn 4.12 that

$$\eta_i = (1 - \eta_i Da)^r \qquad (4.15)$$

where the Damköhler number for crystal growth, which represents the ratio of the pseudo-first-order rate coefficient at the bulk conditions to the mass transfer coefficient, is defined by

$$Da = k_r(c_\infty - c_{eq})^{r-1}/k_d. \qquad (4.16)$$

This relation is plotted in Fig. 4.16.

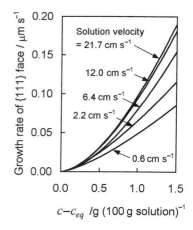

Fig. 4.15 The growth rate of the {111} faces of potash alum crystals increases with both supersaturation and solution velocity.

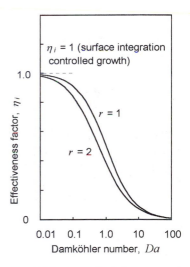

Fig. 4.16 When Da is large, growth is diffusion controlled and $\eta_i \to Da^{-1}$. Conversely, when Da is small, $\eta_i \to 1$ and growth is controlled by the integration step.

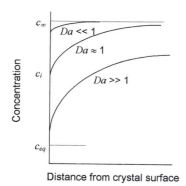

Fig. 4.17   Concentration profiles adjacent to a growing crystal surface.

If the integration step has a much higher rate constant than the mass transport step (i.e. if the Damköhler number $Da \gg 1$) the overall rate of growth is controlled by the mass transfer process. In this case, the driving force for the mass transfer step must be large to force the rate of the transport step to match that of integration. As a result the concentration of solute at the crystal surface, c_i, is close to the equilibrium value, c_{eq}, as shown in Fig. 4.17. The driving force for the transport step is essentially equal to $(c_\infty - c_{eq})$ while the driving force for the integration step is very small. If the rate constants for the two steps are comparable (i.e. $Da \approx 1$) they both participate in determining the overall rate of growth while if $Da \ll 1$ growth, is surface integration controlled.

Temperature usually has a profound effect on the crystal growth rate and effectiveness factors can be used to demonstrate this. The temperature dependence of both the diffusion and integration steps can be characterized by activation energies:

$$k_d = k_{do} \exp(-\Delta E_d/RT) \tag{4.17}$$
$$k_r = k_{ro} \exp(-\Delta E_r/RT). \tag{4.18}$$

ΔE_d is typically in the range 8 to 20 kJ/mol while values of ΔE_r are invariably higher, typically 40 to 80 kJ/mol. As a result the rate of the integration process increases much more rapidly with temperature than does the diffusion process and so crystal growth rates tend to be diffusion controlled at high temperature and integration controlled at low temperatures. Values of η_i are thus close to unity at low temperatures but much lower values are obtained at high temperatures. This is illustrated in Fig. 4.18.

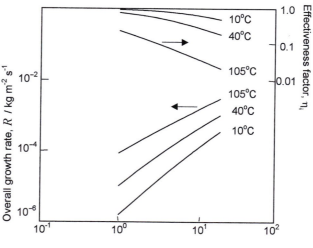

Fig. 4.18   Effect of temperature on growth rate.

References

Burton W. K., Cabrera N. and Frank F. C. (1951) *Philos. Trans. Roy. Soc.* **243** 299.

Davey R. J., Ristic R. I. and Zizic B. (1979) *J. Cryst. Growth* **47** 1.

Davey R. J., Milisavljevic B. and Bourne J. R. (1988) *J. Phys. Chem.* **92** 2032.

Donohoe B., Price R., Taylor G. L. and Halfpenny P. J. (1999) Department of Pure and Applied Chemistry, University of Strathcylde, U.K.

Frank F. C. (1949) *Discuss. Faraday. Soc.* **5** 48.

Garside J. (1991) in *Advances in industrial crystallization*, Garside J. *et al.*, (eds), p. 92, Butterworth-Heinemann, Oxford.

Levenspiel O. (1972) *Chemical reaction engineering*, 2nd edn, Wiley, New York.

Temkin D. E. (1966) in *Crystallization processes*, p. 15, Consultants Bureau, New York.

Further reading

Hartman P. (ed.) (1973) *Crystal growth: an introduction*, North Holland, Amsterdam.

Mullin J. W. (1992) *Crystallization*, 3rd edn, Butterworth-Heinemann, Oxford.

Ohara M. and Reid R. C. (1973) *Modeling crystal growth rates from solution*, Prentice-Hall, New York.

5 Crystal morphology

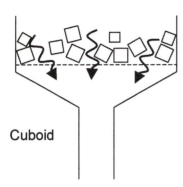

Cuboid

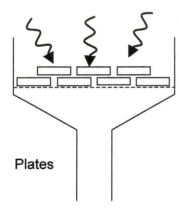

Plates

Fig. 5.1 In many chemical processes the product is isolated by crystallization. If these crystals are cubic in shape then they will filter and wash easily. If they are leaf- or plate-like then they will tend to pack as an impervious layer and will not filter or wash.

5.1 Introduction

The overall shape of a crystal, often called its habit, form or morphology, is a vital parameter in determining the viability of both processes and products. It is usual to distinguish between 'equilibrium' and 'growth' morphologies. The former is the shape that a crystal adopts when allowed to equilibrate with its surroundings and corresponds to the minimization of the surface free energy of the crystal, while the latter is the shape that a crystal develops during growth when kinetics may dominate.

Morphology is determined by two factors, the symmetry of the internal crystal structure, which is manifest in the point group symmetry of the crystal form, and the relative growth rates of the faces bounding the crystal, which are related to the energetics of molecule or ion attachment to the crystal surfaces. The surface-specific nature of this attachment process means that crystal morphology can be dramatically influenced by external factors such as the level of supersaturation, temperature, the growth solvent, and solution purity.

5.2 Predicting crystal morphology

If the crystal structure of a material is known then its crystal morphology may, in principle, be predicted. Structural information provides the necessary symmetry constraints (the point group) together with the positional coordinates of all the atoms in the structure. As shown in Chapter 4, the relative growth rates of the crystal faces can be assessed by calculating the strength of bonds formed when a growth unit joins different surfaces. This simple concept can be extended by making two assumptions:

(a)   A crystal surface, designated by the Miller indices (hkl), grows by the addition of complete layers having the same crystal structure as the bulk and a thickness d_{hkl}, the lattice spacing in the direction perpendicular to the face.

(b)   The energy released per mole of layer added to a crystal face, the so-called *attachment energy* E_{att}, is directly proportional to the growth rate of a given face.

The introduction and use of these approximations is attributed to the pioneering work of Hartman and Perdok (1955). In this way the problem of predicting morphology reduces to the calculation of E_{att} for various crystal faces. The faces that bound the crystal will be those with the lowest values of E_{att} that are consistent with the point group symmetry. E_{att} can be calculated using various approximations.

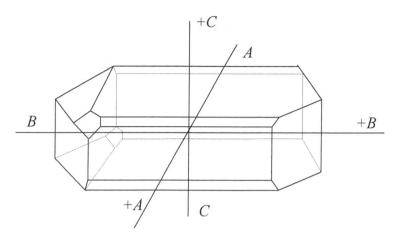

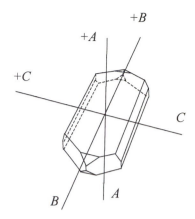

Fig. 5.3 The habit of a sucrose crystal which has no mirror plane perpendicular to the rotation axis and hence has no centre of symmetry. Its point group is 2. Sucrose is an example of a *polar morphology* reflecting the chiral nature of the molecule that is also asymmetric.

Bravais–Friedel–Donnay–Harker

This approach assumes that the binding energy between crystal planes is inversely proportional to the interplanar spacing, i.e. the closer the molecules, the larger are their interaction energies. From this it follows that the growth rate v_{hkl} of the (hkl) face is proportional to the inverse of the distance between the (hkl) planes in the crystal structure, $1/d_{hkl}$. Thus, the relative growth rates of a series of faces can be assessed purely on the basis of the known crystal structure.

Specific force fields (molecular mechanics)

With increasing availability of cheap computer power, the precise calculation of E_{att} using available potential functions to describe the interactions between molecules and ions is now possible (Clydesdale *et al.*, 1993).

The interaction energies E_{ij} between the individual *i*th and *j*th non-bonded atoms of the molecules that constitute the crystal are usually calculated in terms of van der Waals and electrostatic contributions through an equation of the form:

$$E_{ij} = \frac{-A}{r_{ij}^6} + \frac{B}{r_{ij}^{12}} + \frac{q_i q_j}{\varepsilon r_{ij}} \qquad (5.1)$$

in which the van der Waals energy is expressed in terms of the well-known 6–12 Lennard-Jones function and the electrostatic term as a function of the fractional charges q on the *i*th and *j*th atoms, their separation r_{ij} which is known from the crystal structure, and the dielectric constant ε. The example of succinic acid shows the outcome of such calculations with the values of E_{att} for various faces shown in Table 5.1.

Using these as a measure of the relative growth rates of the faces together with the known point group symmetry, 2/m, the growth morphology may be

Fig. 5.2 The morphology of a saccharin crystal grown from ethanol solution. The symmetry elements of this morphology are a twofold rotation axis and a mirror plane. Conventional nomenclature denotes a rotation axis by its numerical order and a mirror plane by the letter m. The point group symmetry of this crystal is 2/m (two upon m), the / signifying that the mirror plane and the rotation axis are at right angles to each other.

It was Auguste Bravais (1811–1863), a typical 19th century scientific 'all-rounder', who first attempted to quantify the link between morphology and structure. The crystallographer George Friedel (1865–1933) continued this work and was visited in Strasbourg, in 1931, by the young American mineralogist, Donnay, who consequently developed an interest in problems of crystal morphology. Donnay's later collaboration with the crystallographer David Harker led in 1937 to further generalization of Bravais' ideas.

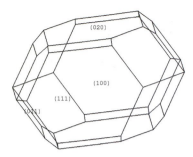

Fig. 5.4 Succinic acid is a simple dicarboxylic acid. Here we show the predicted morphology which corresponds to the observed morphology of crystals grown from the vapour.

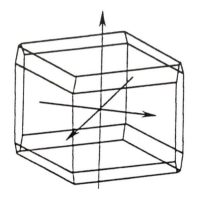

Fig. 5.5 The predicted morphology of barium sulphate.

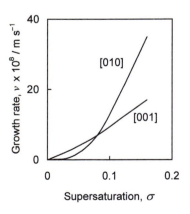

Fig. 5.6 Effect of supersaturation on the growth kinetics of paracetamol crystals (after Finnie *et al.*, 1996).

Table 5.1 Attachment energies for succinic acid

Face {hkl}	{020}	{111}	{100}	{110}	{011}	{021}
E_{att}/kJ mol^{-1}	−51.8	−63.1	−58.3	−73.6	−72.4	−70.4

predicted and is shown in Fig. 5.4. It is worth noting that the {100} and {020} faces that grow most slowly occupy the largest areas in the final morphology. The observed morphology of crystals grown from the vapour shows that the crystals are thinner but essentially similar to the calculated habit.

Atomistic lattice simulations

In the case of ionic crystals, energetic parameters may again be calculated by use of potential functions. These allow for both long-range electrostatic forces and short-range interactions and assign appropriate charges to the constituents of polyatomic species. In this methodology, it has been conventional to calculate not attachment energies but surface energies directly (Mackrodt *et al.*, 1987). Thus, it is imagined that a crystal is cleaved to expose the desired face, (hkl). The energy changes involved are calculated so that the surface energy is given by:

$$\gamma_{hkl} = 0.5(E_{hkl}^{B} - 2E_{hkl}^{S}) \tag{5.2}$$

where the energy of the crystal plane (hkl) is E_{hkl}^{B} in the bulk and E_{hkl}^{S} in the surface. In the case of barium sulphate the calculated values are given in Table 5.2.

Table 5.2 Predicted surface energies of barium sulphate crystals

Face {hkl}	{001}	{010}	{100}	{011}	{101}	{210}	{211}
γ_{hkl}/J m^{-2}	0.63	0.96	1.30	1.59	1.10	0.71	0.84

The morphology predicted from the values in Table 5.2 is shown in Fig. 5.5. Because these morphologies are based on surface energies they correspond to the equilibrium morphology while the other two methods described yield growth morphologies.

5.3 The influence of supersaturation, temperature, solvents, additives and impurities

The predictions discussed above assume that structural factors alone control crystal growth rates. It is known, however, that the relative growth rates of different crystal surfaces can be particularly susceptible to the conditions prevailing in the growth medium. We have seen in Chapter 4 how the growth rate of a face can depend in a non-linear way on supersaturation. This makes it possible for growth kinetic curves to intersect as shown in Fig. 5.6. This shows the growth rates of paracetamol crystals in the [010] and [001] directions when grown from aqueous solution (Finnie *et al.*, 1996). Because the growth rate curves for these two faces intersect, needles elongated along [001] are produced at low supersaturations while at high supersaturations the crystals are bipyramidal.

As discussed in Chapter 4 (Eqns 4.17 and 4.18) crystal growth is a thermally activated process in which growth rates increase with temperature (as do chemical reactions) according to an Arrhenius relationship. One consequence of this is that growth tends to become diffusion controlled with increasing temperature. This leads to the general rule of thumb that crystals are more isotropic in their shape when grown at higher temperatures.

In addition to these global kinetic effects, specific morphological changes can occur in both solution and melt growth due to the selective adsorption of non-crystallizing components on particular crystal surfaces. For example, the presence of specific impurities even at low (ppm) levels can have a profound effect on crystal morphology. The converse of this is that impurities are sometimes deliberately added to produce a desired morphological effect; in this case they are referred to as *additives*.

When impurities become adsorbed at a growing surface so strongly that they block kink sites then the growth rate of the face is reduced. Because the concentration of active growth sites on the surface is low, additives and impurities can exert significant effects at very low concentrations. Similar effects are seen in the poisoning of catalysts. The result of such surface-specific adsorption processes can often be to bring about a change in crystal morphology. This is demonstrated in Fig. 5.7 where specific adsorption on the fastest growing faces modifies a needle morphology to a more equant shape. When designing an additive to inhibit crystallization we work in the reverse way; the additive should be targeted to adsorb selectively on the fast growing faces. Figure 5.8 illustrates schematically how impurities adsorbed at a step can block growth.

The reduction in crystal growth rate (v_i in impure and v_∞ in pure solution) can be expressed by an equation of the form:

$$v_i = v_\infty(1 - 2r_c/l)^{0.5}. \tag{5.3}$$

As the separation l between adsorbed impurities on the surface decreases towards the dimension of the critical radius r_c of a two-dimensional nucleus, steps cannot advance and growth eventually stops. Figure 5.9 displays kinetic data for three molecular systems showing how the growth rate falls rapidly with increasing additive concentration.

Solvents may act in a similar way but they also determine the value of α (see Section 4.2) for a particular interface. It follows that for growth from solution, morphological changes can result not only from strong solvent adsorption but also from surface-specific changes in growth mechanisms

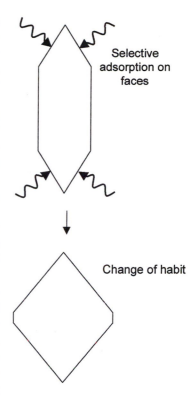

Selective adsorption on faces

Change of habit

Fig. 5.7 Specific adsorption onto the fast growing faces of a needle form changes the morphology to give a more equant shape.

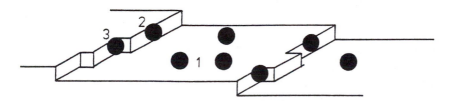

Fig. 5.8 Adsorption of impurities (1) on a step, (2) at a ledge, and (3) at a kink site.

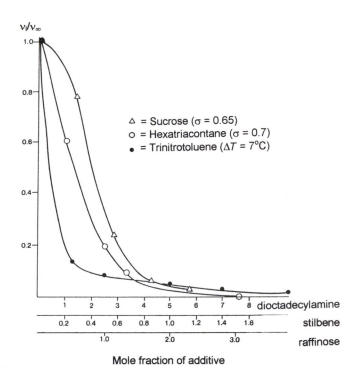

Fig. 5.9   Kinetic data showing the decrease in growth rate with increasing additive concentration (after Black *et al.*, 1986).

brought about by solvent changes. Since α is related to solubility by the relation

$$\alpha = \xi\big[(\Delta H_f/RT) - \ln x_{eq}\big], \tag{4.4}$$

increasing solubility is seen to decrease α and cause growth rates to increase. Growth of the {110} faces of hexamethylene tetramine provides such an example (Fig. 5.10). When grown from the vapour, from ethanol and from water, and hence as the solubility increases, the growth mechanism changes from one that is controlled by screw dislocations, through surface nucleation to continuous growth.

5.4   Some specific examples

It is apparent that particular morphological effects are largely structurally driven and for this reason are specific to each crystallizing system. In this section we will look at some examples chosen to demonstrate the effects discussed above.

The influence of solvent: succinic acid

The predicted morphology of succinic acid and its good agreement with that observed for crystals grown from the vapour has been illustrated in Fig. 5.4.

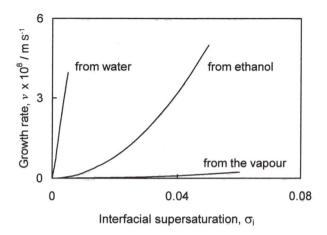

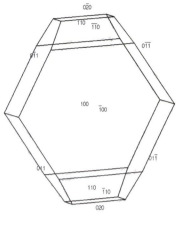

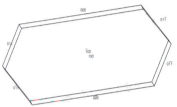

Fig. 5.10 Dependence of the {110} face growth rate of hexamethylene tetramine on solvent.

Crystals grown from aqueous and alcoholic solutions, however, have a different morphology as is evident from Fig. 5.11. Whereas the {020} faces dominate in vapour growth, the {100} faces dominate when crystallization takes place from a solvent. By visualizing the molecular packing in the various crystal surfaces we can gain further insight into this phenomenon. This may be done by referring back to the crystallographic projection shown in Fig. 4.5 since succinic acid, $COOH(CH_2)_3COOH$, has essentially the same crystal structure as adipic acid. For the {100} surface the polar carboxylic acids groups protrude into the solution where they will interact strongly with a polar solvent. This may be contrasted with the {020} faces which, like the {001} face of adipic acid, lie parallel to the alkyl chains. This preferential solvent binding results in the {100} faces having slower growth rates and hence being more important faces than expected from the prediction. The kinetics of crystal growth shown in Fig. 5.12 reflect this with the {020} growing faster in both water and isopropanol.

These kinetic data also reveal that at equivalent supersaturations crystal growth is slower from isopropanol than from water. Values of the α-factors for each face are shown in Table 5.3. These show no significant effect of solvent on the value of α and hence on the nature of the interface. Thus, no change in growth mechanism would be expected, implying that the growth rates in water are likely to be the same as those in isopropanol. The observed lower growth rates in isopropanol can only be explained by a specific stereochemical effect related to solvent adsorption, presumably linked to hydrogen bonding of water or isopropanol with surface carboxylic acid groups. The larger size of the

Fig. 5.11 The morphology of succinic acid crystals grown from (a) water (above) and (b) isopropanol (below). What is their point group symmetry?

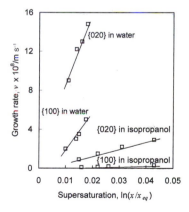

Fig. 5.12 The growth kinetics of succinic acid (Davey *et al.*, 1982).

Table 5.3 α-Factors for succinic acid in water and in isopropanol

Solvent	(020) face	(100) face
Water	7.8	5.8
Isopropanol	8.0	6.0

isopropanol molecule would make it a more effective inhibitor of growth and hence give rise to the observed relative growth rates in water and isopropanol.

The influence of impurities—urea and biuret

One of the first recorded examples of an impurity effect is that noted by Rome de l'Isle in 1783. Common salt crystallized in the presence of horse urine was found to produce octahedral rather than cubic salt.

When urea [$(NH_2)_2CO$] is synthesized from ammonia and carbon dioxide a dimer, biuret ($NH_2CONHCONH_2$) is also formed in small amounts. Urea crystals grown in pure aqueous solutions form long [001] needles while in solutions that also contain biuret the crystals are shorter and stubby, having had growth in [001] suppressed (Davey *et al.*, 1986). The structural basis for the needle-like morphology of urea lies in the strong intermolecular hydrogen bonding along the c-axis as shown by the dotted lines in Fig. 5.13. Furthermore, the structure is such that the {001} surfaces cannot easily discriminate between two urea molecules and one biuret molecule. The biuret molecules can therefore become attached to the lattice at growth sites in the [001] direction. For urea molecules trying to bind to a biuret-contaminated surface, however, life is difficult since the NH_2- groups in the crystal surface that are needed to form hydrogen bonds are now missing. This effectively reduces the growth rate in the [001] direction, so resulting in stubby crystals.

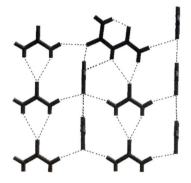

This example illustrates the important general rule that additives capable of modifying crystal morphologies are those that are in some way molecular impostors. They are able to enter the growing surface and yet once there, they disrupt further growth. To perform this function effectively the additive molecule must resemble the crystallizing molecule while containing some small difference in stereochemistry or functionality, rendering it capable of inhibiting growth in a selected direction. In recent years this principle of *tailor-made additives* has been applied to the rational design of crystallization inhibitors for a number of applications (Davey *et al.*, 1992; Weissbuch *et al.*, 1994).

Fig. 5.13 Urea is widely used as a fertiliser. In the reaction $2NH_3 + CO_2 \rightarrow NH_2CONH_2 + H_2O$ an impurity, biuret, is also formed. Biuret, a dimer of urea, kills citrus fruits and pineapples and so must be removed from the product by crystallization. The figure shows a biuret molecule occupying two urea sites at the fast-growing (001) face.

5.5   Crystal purity

In many industrial situations crystals are grown from solvents under impure conditions. As we have seen, reaction by-products such as biuret may themselves act as tailor-made additives for the growth of the crystallizing material. In some applications such as agrochemicals, pharmaceuticals, polymer intermediates and electronic materials, crystallization is used as a means of purification. This relies on the selectivity of the crystal-liquid interface and its power in discriminating between various solute species. There is clearly a link between purification and habit modification by impurities, since those molecules that bind selectively to surfaces are also those most likely to be incorporated into the crystal lattice.

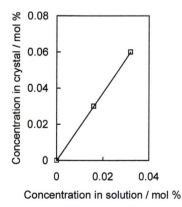

The degree of purification by crystallization is usually described quantitatively by a distribution segregation coefficient defined as:

$$D_i = \frac{\text{mass of impurity per mass of solute in the crystal}}{\text{mass of impurity per mass of solute in the solution}}.$$

Fig. 5.14 The incorporation of octanoic acid into adipic acid crystals (Davey *et al.*, 1992).

If $D_i > 1$ the relative concentration of the impurity in the crystal is higher than in the solution and crystallization would lead to enrichment of the impurity in the solid phase. Only if $D_i < 1$ does crystallization result in purification. Figure

5.14 shows data for the segregation of octanoic acid between solid and aqueous solutions of adipic acid. The slope of this plot indicates that the value of D_i is approximately 2 suggesting that octanoic acid, a crystallization inhibitor of adipic acid, segregates preferentially into the growing crystals.

In general, the extent of incorporation of an impurity or additive will depend on the relative size, stereochemistry and functionality of the additive molecule when compared with the crystallizing solute. The more dissimilar the components, the purer the crystals are likely to be.

References

Black S. N., Davey R. J. and Halcrow M. (1986) *J. Cryst. Growth* **79** 765.

Clydesdale G. , Docherty R. and Roberts K. J. (1993) in *Colloid and surface engineering: controlled particle, droplet and bubble formation,* D. Wedlock (ed.), Butterworth-Heineman, Oxford.

Davey R. J., Mullin J. W. and Whiting M. J. L. (1982) *J. Cryst. Growth* **58** 304.

Davey R. J., Fila W. and Garside J. (1986) *J. Cryst. Growth* **79** 607.

Davey R. J., Polywka L. A. and Maginn S. J. (1991) in *Advances in industrial crystallization*, J. Garside, R. J. Davey, A. G. Jones (eds), p. 150, Butterworth-Heinemann, Oxford.

Davey R. J., Black S. N., Logan D., Maginn S. J., Fairbrother J. E. and Grant D. J. W. (1992) *J. Chem. Soc. Faraday Trans.* **88** 3461.

Finnie S., Ristic R. I., Sherwood J. N. and Zikic A. M. (1996) *Trans. IChemE* **74** 835.

Hartman P. and Perdok W. (1955) *Acta Cryst.* **8** 49 and 521.

Mackrodt W. C., Davey R. J. , Black S. N. and Docherty R. (1987) *J. Cryst. Growth* **80** 441.

Weissbuch, Popvitz-Biro R., Leiserowitz L. and Lahav M. (1994) in *The lock and key principle,* J. P. Behr (ed.), p. 173, Wiley, New York.

Further reading

General:

Mullin J. W. (1992) *Crystallization*, 3rd edn, Butterworth-Heinemann, Oxford.

Further details of solvent effects are given in:

Davey R. J. (1982) in *Current topics in materials science*, Vol. 8, E. Kaldis (ed.), p. 431, North Holland, Amsterdam.

6 Polymorphism

6.1 Introduction

The three polymorphs of calcium carbonate are calcite, aragonite and vaterite. They differ in the arrangement of calcium and carbonate ions and their solubility products at 298 K are 3.31×10^{-9}, 4.61×10^{-9} and 1.22×10^{-8} mol^2 dm^{-6} respectively.

We have seen in Chapters 2 and 3 that some molecules exhibit polymorphism, that is they are able to adopt more than one crystal structure. When preparing materials by crystallization it is important to recognize and to be able to control this phenomenon because each polymorph has its own unique combination of mechanical, thermal and physical properties. The production of specific and well-defined polymorphs is therefore crucial in chemical manufacture.

A definition of polymorphic forms was first made by Mitscherlich in 1821 in relation to arsenates, phosphates and sulphur. In 1832 Wöhler and Liebig discovered the first example of polymorphism in an organic material, benzamide. In 1899, Ostwald concluded that almost every substance could exist in two or more solid phases provided the experimental conditions are suitable. The development of solid state chemistry during the twentieth century has confirmed and exemplified this statement. Table 6.1 gives some common examples.

The famous American chemical analyst Walter C. McCrone, who was one of the few scientists entrusted to examine the Turin Shroud, commented in 1963 that virtually all compounds are polymorphic; the number of polymorphs of a material is in direct proportion to the time and money spent looking for them.

In the chemical and pharmaceutical industry the demand for high yields and high production rates has forced chemists and engineers to operate processes far from equilibrium, so exacerbating the tendency to form polymorphic structures that, for a given temperature, pressure and composition, are not the most stable. Such unstable structures will eventually undergo phase transitions to more stable phases. This transformation process and its structural and thermodynamic basis are the subjects of this chapter. Although we only deal specifically with polymorphs it is worth pointing out that materials may also adopt other different solid forms such as:

Solvates

Gypsum, the major component of wall plaster, is a calcium sulphate dihydrate; citric acid, the lemon flavouring used in foods, can crystallize as a monohydrate.

In this case both solute and solvent molecule are part of the crystal structure that make up the solid phase. This results in a crystal of defined composition in which the solvent molecules adopt well-defined lattice positions.

Compounds

In multicomponent systems, particularly aqueous solutions of ionic materials, the composition of the solid phase that is crystallized can depend on composition. In the system $MgCl_2$–KCl–H_2O, crystals of a compound salt $KCl \cdot MgCl_2 \cdot 6H_2O$ can precipitate, while in the $NaCl$–urea–H_2O system the compound $NaCl \cdot Urea \cdot H_2O$ can form.

Amorphous solids

The term amorphous is usually used in relation to an X-ray diffractogram that shows no diffraction peaks and hence no structural periodicity or long-range order.

These are formed in both ionic and molecular systems and are solid phases in which there is no long-range order. Common window glass is perhaps the best known example, being a mixture of metal oxides, but amorphous forms of calcium phosphate and aluminium phosphate and sulphates are also known. Amorphous ice results when water is frozen very rapidly, while sugars are known to form glassy solids.

Mesophases

Occasionally, when materials precipitate at very high supersaturations or when they comprise molecules that exhibit amphiphilic behaviour, they appear not as solids but as liquid-like droplets. Such droplets may be undercooled isotropic liquid or they may be liquid crystalline in nature. In the latter case the molecules will be ordered in one or two dimensions and the drops will be birefringent when observed in the polarizing microscope.

Amphiphilic molecules have well-defined polar and non-polar domains. Surfactants are good examples; the major component of washing-up liquids forms a liquid crystalline lamella phase as well as a crystalline phase.

Table 6.1 Examples of common chemical products that are polymorphic

Chemical product	Number of polymorphs	Applications
Ammonium nitrate	5	Explosives, fertilizers
Aspirin	4	Pharmaceutical
Calcium carbonate	3	Filler for plastics
Copper phthalocyanine	4	Pigment
Indigo	2	Disperse dye
Lead azide	2	Explosive
Lead chromate	2	Pigment
Pentaerythritol tetranitrate	2	Explosive
Phenobarbitone	13	Pharmaceutical
Sorbitol	2	Sweetener
Sulphathiazole	4	Pharmaceutical
Tetraethyl lead	6	Fuel additive
Titanium dioxide	3	Pigment
Titanium trichloride	4	Catalyst
Triglycerides	4	Fats and oils

6.2 Thermodynamics

Two polymorphic modifications of a given material constitute two homogeneous phases. As we saw in Chapter 2, if they are in equilibrium with each other they must conform to the Gibbs phase rule:

$$F = C - P + 2.$$

With $C = 1$ and $P = 2$ the system has only one degree of freedom. Consequently, at constant pressure, equilibrium between the polymorphs occurs at a fixed temperature, while at constant temperature it occurs at a fixed pressure. In addition, since F cannot be negative, no more than three polymorphs can be in equilibrium with each other.

There is no universal convention for naming polymorphs; sometimes Greek letters α, β, γ are used, sometimes the Roman numerals I, II, III, etc. It is important to note that the latter do not imply anything about the relative stability of the various forms, rather they usually follow the order of discovery.

Temperatures at which two or more polymorphs are in equilibrium are called *transition temperatures* and their dependence on pressure is given by the Clapeyron equation:

$$\frac{dT}{dP} = \frac{T_t \Delta V}{\Delta H_t}. \tag{6.1}$$

The painkiller paracetamol is polymorphic. The stable form I crystallizes from aqueous solutions whilst the metastable form can only be grown from ethanol. Such solvent dependence of polymorph appearance is common, but is related to kinetics not thermodynamics.

ΔV is the difference in molar volume of the polymorphs, ΔH_t the latent heat of the transition and T_t the transition temperature at a given pressure. This equation makes it possible to calculate the change ΔT in the transition temperature due to a change in pressure ΔP. We have seen in Chapter 2 that for a crystallization process the solubility curves provide important data. In a dimorphic system consisting of polymorphs I and II, with polymorph II being the more stable at the temperature of interest, it follows that for molecules in the solid state:

$$\mu_{solid}(II) < \mu_{solid}(I). \qquad (6.2)$$

If we place a crystal of each polymorph in contact with its saturated solution, then the chemical potentials of molecules in solid and liquid phases are equal. We can therefore write each in terms of the solubility, x_{eq}, of each phase in a given solvent. Thus assuming ideal solutions:

$$\mu_{solid}(II) = \mu_{eq}(II) = \mu^0 + RT \ln x_{eq}(II)$$

and

$$\mu_{solid}(I) = \mu_{eq}(I) = \mu^0 + RT \ln x_{eq}(I).$$

Combining these relations with the condition given by Eqn 6.2 shows that

$$x_{eq}(II) < x_{eq}(I). \qquad (6.3)$$

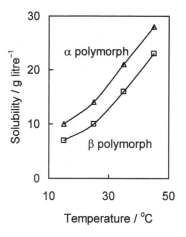

Fig. 6.1 L-Glutamic acid is an amino acid constituent of soy sauce and famed amongst lovers of Chinese food as the organic anion in monosodium glutamate (MSG). It has two polymorphs α and β. The solubility curves are shown here— which is the more stable structure?

Hence the relative solubility of each polymorph reflects its relative stability, the more stable polymorph having the lower solubility. This relationship is completely independent of the solvent.

In polymorphic systems phase diagrams (i.e. solubility curves) are found to fall into one of the two categories seen in Fig. 6.2:

Monotropic in which the relative solubilities of the polymorphs are independent of temperature, and

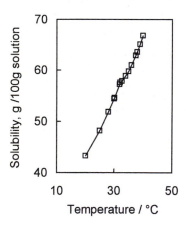

Fig. 6.3 Ammonium nitrate is a fertilizer and an explosive. Its aqueous solubility is shown here. Is it an enantiotropic or monotropic system?

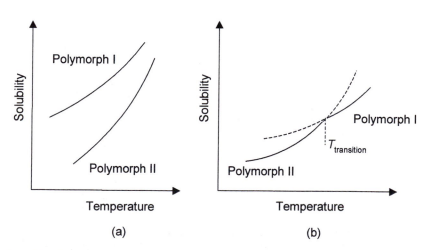

Fig. 6.2 Solubility curves in (a) monotropic and (b) enantiomorphic systems.

Enantiotropic in which the relative solubility of the polymorphs is temperature dependent.

In enantiotropic systems the transition temperatures are revealed as discontinuities where the individual solubility curves intersect (Fig. 6.2b).

6.3 Structure

The existence of polymorphic structures is a reflection of alternative ways in which molecules or ions can pack in a crystal in attempting to minimize their free energy whilst crystallizing at a reasonable rate. In some cases different packing is driven by intermolecular interactions (enthalpy) and in others by considerations of entropy. Figure 6.4 shows three different packing motifs for *n*-alkanes. With other molecules, for example L-glutamic acid shown in Fig. 6.5, not only are the packing sequences of the polymorphs different but so too are the molecular conformations.

6.4 Phase transitions

Examination of phase diagrams such as those in Fig. 6.2 shows that it is possible to prepare metastable crystalline solids either by rapid crystallization in a monotropic system or by temperature change in an enantiotropic system. The formation of such a metastable state will be followed by transformation to the most stable one as the system strives to reach equilibrium.

In the transformation process molecules in the solid phase undergo appropriate positional changes to yield a new structure with lower free energy. One commonly used method of characterizing such transitions at constant pressure is to consider how various thermodynamic functions change at the transition temperature. An *n*th-order transition is defined as one for which the *n*th derivative of the free energy with temperature is discontinuous. A *first-*

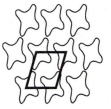

Hexagonal

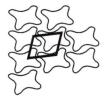

Oblique

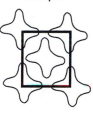

Cubic

Fig. 6.4 Packing modes of alkane molecules viewed end-on.

In 1897 Ostwald published his 'Rule of Stages'. This was a statement of his experimental observations that when systems crystallized they often formed metastable phases first and then transformed in stages, each transformation lowering the free energy of the system until it reached equilibrium. Fig. 6.8 shows this for potassium nitrate.

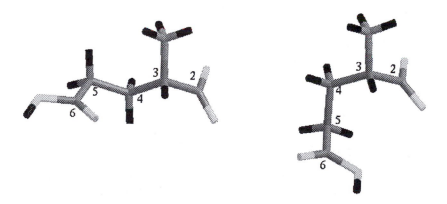

Fig. 6.5 The conformations of glutamic acid in its two polymorphs. Note how the configuration at carbon atoms 2 and 3 is identical. The conformations (β left, α right) are interconverted by rotation at carbons 3 and 4.

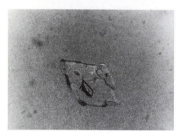

Fig. 6.6 The change in morphology resulting from the solid state transformation between forms I (above) and II (below) of terephthalic acid (after Davey *et al.*, 1993). The crystals are about 100 μm in size.

Pure terephthalic acid (PTA) is a key intermediate in the production of polyester. It is purified by crystallization on the scale of 10^6 tonnes p.a. before it is reacted with ethylene glycol to form the polymer. Since this is a solid–liquid reaction, the crystal morphology of the product is crucial in ensuring that it can be easily mixed with the ethylene glycol. Commercial PTA is sold as the metastable polymorph I. It is fortunate that the transformation to the stable form is kinetically hindered at room temperature.

order transition is characterized by discontinuous entropy, enthalpy and heat capacity curves whereas a *second-order transition* is characterized by continuous entropy and enthalpy but discontinuous heat capacity curves.

Structurally, the changes involved can be defined precisely if the crystal structures of each phase are known. In the absence of routine crystal structure determination, physicists have in the past used a number of generic terms that remain in use.

Reconstructive transformations

Here the two structures are so different that the transition between them can only be affected by the disintegration of one and reconstruction of the new. A well-known example is the aragonite to calcite transformation in calcium carbonate. It is common in systems of this type that the transformation is catalysed if the crystals are in contact with their saturated solutions, since dissolution of the metastable structure is a convenient way of disintegrating it.

Displacive transformations

In this case, molecules of one structure can be displaced to yield the new structure in such a way that nearest neighbour interactions are preserved and only second neighbours change. In these cases, activation energies are low, transformation rates are high and the process can be accomplished in the solid state. An example of this class is the I to II transition in terephthalic acid in which hydrogen-bonded sheets of one structure slide and twist to give the new structure and the transformation can proceed without disrupting the parent crystal. Figure 6.6 shows this process as evidenced by the morphological change.

Since the free energy change associated with a transformation can be written as

$$\Delta G = \Delta H - T\Delta S, \qquad (6.4)$$

it follows that a transformation can be accompanied by either an exothermic or an endothermic enthalpy change as long as the overall ΔG is negative. If the enthalpy change is endothermic then it must be balanced by a positive change in entropy and hence the new phase is more disordered than its precursor. Such a transition is termed an *order-disorder transformation* and is again exemplified by the terephthalic acid transition.

Differential scanning calorimetry (DSC) is a convenient means of monitoring phase transformations and measuring associated enthalpy changes. Figure 6.7 shows such a trace for a synthetic dyestuff. A phase transition, followed by melting, is clearly seen. The superimposed x-ray diffraction (XRD) powder patterns indicate that a structural change is involved.

6.5 Kinetics

The kinetic processes involved in phase transitions between polymorphs depend largely on the extent of structural changes involved. Following the above discussion, two mechanistic scenarios can be described.

Solvent-mediated (reconstructive) transformation, in which the metastable phase dissolves while the stable phase renucleates and grows from solution. Figure 6.8 is an example of this. This route often involves a lower activation energy than a solid state process and hence is favoured for reconstructive transitions occurring well below the melting point.

Solid state (displacive) transformation, in which nucleation and growth of the new phase take place in crystals of the unstable phase. Such transitions are often reversible (enantiotropic) when the temperature is raised and lowered through the transition temperature.

For solid state transformations, Fig. 6.9 illustrates the basis of the derivation of kinetic expressions (Cardew *et al.*, 1984). Nucleation is assumed either to be instantaneous or to take place at a constant rate during the process. The relationship between the transformed volume *V* and time is then given by:

$$V(t)/V(total) = 1 - \exp(-kt^n). \qquad (6.5)$$

The value of *n* is determined by the dimensionality (1, 2 or 3) of the growth process.

The consequence of reducing large crystals to a powder mirrors the dispersion of a liquid into drops as illustrated in Fig. 3.12—the transformation can become nucleation limited because some fragments will no longer contain nuclei. In this case, the kinetic expression changes to:

$$V(t)/V(total) = 1 - \exp(-t/t_n) \qquad (6.6)$$

in which t_n is the nucleation time. Figure 6.10 illustrates kinetic data for the IV to III transition in ammonium nitrate powders; the phase diagram for this system is given in Fig. 6.3. The characteristic S-shape curves are evident and the data at 35°C have been fitted to both of the above equations.

For solvent - mediated transformations the extent of transformation, *V(t)/V(total)*, does not yield useful mechanistic information (Cardew and Davey, 1985). Consideration of Fig. 3.7 shows that this is because the kinetics of this process are controlled by the relative rates of dissolution of the metastable (I) and growth of the stable (II) phases. Thus, when a metastable phase is in contact with its mother liquor this liquid is supersaturated with respect to the stable phase and hence crystals of form II can also nucleate and grow. Growth

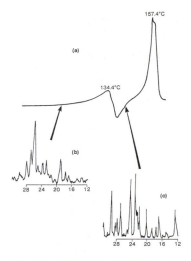

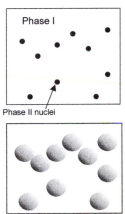

Fig. 6.7 DSC (a) and XRD (b and c) data for a solid state transformation in a dyestuff induced by heating. The transformation takes place at 134.4°C followed by melting at 157.4°C (after Davey and Richards, 1985).

Fig. 6.9 A solid state polymorphic transition. Nuclei of the new structure grow throughout the crystal until they impinge on each other.

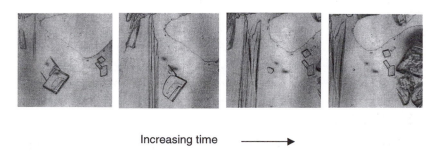

Increasing time ⟶

Fig. 6.8 A solution-mediated transition in potassium nitrate. Note how the rhombic crystal dissolves as the needle grows.

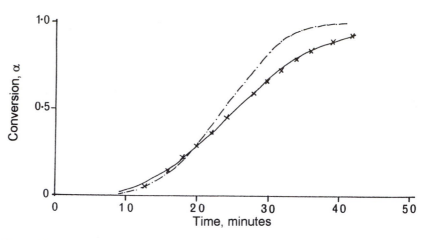

Fig. 6.10 Transformation kinetics in ammonium nitrate (after Cardew *et al.*, 1984). The solid line corresponds to eqn (6.6), crosses to the measured data.

of II reduces the solution composition below the saturation level of form I, causing more form I to dissolve. This process continues until all the form I has dissolved and recrystallization to form II is complete.

Referring to Fig. 3.7, two supersaturations can be defined. The first is the driving force for the growth of polymorph II that is given by:

$$\sigma = (x - x_{II})/x_{II}$$

while the second is the supersaturation at the start of the transformation process:

$$\sigma_{I,II} = (x_I - x_{II})/x_{II}.$$

A kinetic description of this overall process is then given by three simultaneous equations describing:

(i) dissolution of form I:

$$dr_I/dt = -k_D(\sigma_{I,II} - \sigma) \qquad (6.7)$$

(ii) growth of form II:

$$dr_{II}/dt = k_G\sigma \qquad (6.8)$$

(iii) the mass balance:

$$\sigma = \sigma_i - (\sigma_i - \sigma_{I,II})(r_I/r_{I,i})^3 - \sigma_i(r_{II}/r_{II,f})^3. \qquad (6.9)$$

r_I and r_{II} are the crystal sizes of forms I and II respectively, the subscripts i and f refer to initial and final values of relevant parameters and σ_i is the supersaturation that would exist if all the crystals of both forms were dissolved.

Simultaneous solution of these equations allows the supersaturation–time profiles to be predicted for various situations. A typical result is seen in Fig. 6.11a and a number of general features of such curves can be identified. First, a decrease in supersaturation from the maximum value, $\sigma_{I,II}$, occurs as form II

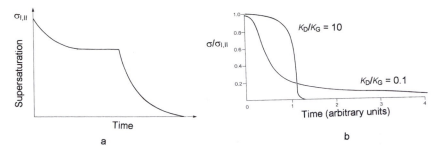

Fig. 6.11 Kinetics of solvent-mediated transitions. (a) General features, (b) dissolution ($k_D/k_G = 0.1$) and growth ($k_D/k_G = 10$) control.

nucleates and grows. Second, there is a period during which the growth and dissolution processes are precisely balanced, resulting in a supersaturation plateau. Finally, when the last of the phase I has dissolved, a reduction in the supersaturation to zero takes place as form II completes its growth. Figure 6.11b shows this in more detail. Desupersaturation profiles have been chosen to illustrate situations in which the transformation is controlled either by growth of form II ($k_D/k_G = 10$) or dissolution of form I ($k_D/k_G = 0.1$). The two extremes are easily distinguishable by the position of the plateau relative to the maximum supersaturation. If the two are close together then form I dissolves rapidly enough to maintain the solution at its maximum allowable composition. The slower the dissolution process, the lower the plateau.

Figure 6.12 shows two examples, one for a dyestuff in which dissolution is rate controlling, and one for an agrochemical in which growth is rate controlling. It is worth noting that because the plateau supersaturation is so low when the dissolution process is controlling, the total transformation time can be very long, perhaps extending to a number of days.

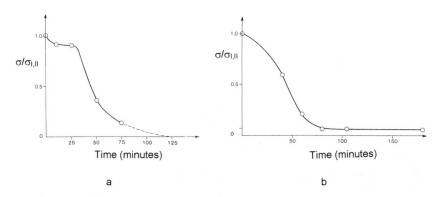

Fig. 6.12 Experimental data for solution-mediated transformations: (a) an agrochemical and (b) a dyestuff (after Davey *et al.*,1986).

6.6   Conclusions

Although polymorphic transformations are structurally driven, mechanistically they show many similarities to other nucleation and growth processes. The implications of these transformation processes are considerable when considering process and product design and development. Crystalline solids may break down to give powders with very different size distributions to the starting material, while solvent-mediated transformations will lead to changes in morphology (or form) and increases in crystal size. In all cases a change in polymorph will impact upon the operation of a process and give a product with variable physical properties.

References

Cardew P. T., Davey R. J. and Ruddick A. J. (1984) *J. Chem. Soc. Faraday Trans.* **80** 659.
Cardew P. T. and Davey R. J. (1985) *Proc. Roy. Soc. Lond.* **A298** 415.
Davey R. J. and Richards J. (1985) *J. Cryst. Growth* **71** 597.
Davey R. J., Cardew P. T., McEwan D. and Sadler D. E. (1986). *J. Cryst. Growth* **79** 648.
Davey R. J. *et al.* (1993) *Nature* **366** 248.
Davey R. J. *et al.* (1994) *J. Chem. Soc. Faraday Trans.* **90** 1003.

Further reading

Verma A. R., Krishna P. (1966) *Polymorphism and polytypism in crystals*, Wiley, New York.
Desiraju G. (1989) *Crystal engineering*, Elsevier, Amsterdam.

7 Number balances and size distribution modelling

7.1 The importance of crystal size distribution

Together with the crystal morphology, the crystal size distribution (CSD) produced within a crystallizer is of crucial importance in determining the ease and efficiency of subsequent solid–liquid separation steps, the suitability of crystals for further processing, their caking and storage characteristics, and the eventual customer appeal of the product. Equally important is the interaction between the CSD and crystallizer operation through the feedback relation illustrated in Fig. 1.2. In particular, the size distribution influences the supersaturation at which the crystallizer operates and so has an effect on the crystal nucleation and growth rates, crystal morphology and purity, fouling of solid surfaces, and the stability of operation.

The most revealing way to explore these interactions is by making use of the *number* or *population balance*. This has been comprehensively described by Randolph and Larson (1988) and is now widely accepted as providing the greatest insight into the description and modelling of crystallizers. Underlying the number balance is the recognition that for any particulate process such as crystallization (other examples would be agglomeration and grinding) a full process description requires that a number balance must be satisfied in addition to the widely recognized mass and energy balances; all the individual particle numbers must be accounted for. It is only by accounting for the conservation of numbers that it is possible to describe the particle size distribution in a theoretical as opposed to an empirical form. Recognizing the necessity for this third conservation equation is at the heart of crystallizer modelling.

A simple crystallization example will make this clear. Suppose that a saturated solution of potash alum is prepared at 80°C. The solution is then cooled to 20°C, crystallization takes place and the system is allowed to come to equilibrium. An energy balance will enable us to calculate the heat duty necessary to perform this cooling operation while a mass balance, in combination with the solubility data, will enable the crystal yield to be determined. These two conservation equations, however, tell us nothing about the size distribution of the potash alum crystal product; we cannot distinguish the case when all the product is produced as a single large crystal from that when the product is in the form of innumerable tiny crystals. It is only by applying the number balance that we can distinguish between these two cases.

Fouling or encrustation is the process by which crystalline material deposits on the walls of the crystallizers. This can cause severe operating problems by reducing heat transfer rates and by reducing the effective volume of the crystallizer. Significant downtime is often needed to remove such deposits.

Population balance models for crystallizers were published in the 1930s and 1940s. Perhaps the most interesting is that of Bransom *et al*. (1949) who described work carried out during World War II on the explosive cyclonite. This was drowned out in a continuous crystallizer by adding water to a solution of the material in concentrated nitric acid. The nucleation and growth kinetics were deduced by application of what we now call the population balance.

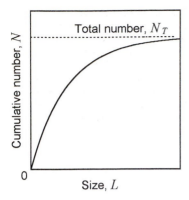

Fig. 7.1 Cumulative number distribution.

7.2 Characterization of size distributions

Since the physical principle behind the derivation of the CSD in a crystallizer is that of a number balance, it is most convenient to develop the basic equations in terms of number distributions. Once these basic equations have been derived, the form of other distributions, for example area and mass, can readily be deduced.

Figure 7.1 represents the cumulative number distribution of a population of crystals. It is often difficult to define the size of a crystal but each crystal is represented here by an average size which can be related to some characteristic of the crystal; for example, to the edge length if the crystal were a cube or to the diameter if it could be approximated by a sphere. The curve represents the number of crystals, N, that are smaller than some crystal size, L, contained within a specific volume of suspension. The asymptote of the curve represents the total number of crystals in the distribution, N_T.

dN crystals are contained within the increment of size dL. The number or population density, n, is defined as:

$$n = \frac{dN}{dL}. \tag{7.1}$$

n is thus the slope of the $N(L)$ curve and represents the number of crystals per increment in size; it has dimensions of number per unit length per unit volume, e.g. number μm^{-1} m^{-3}. For the cumulative distribution $N(L)$ shown in Fig. 7.1, the corresponding $n(L)$ distribution has the form shown in Fig. 7.2.

We will see below that application of the number balance enables us to determine $n(L)$ for a particular crystallizer configuration. Once $n(L)$ is known, the particle numbers within various size ranges can be determined. Thus

$$\text{number of crystals } dN \text{ in the size range } dL = n \, dL \tag{7.2}$$

$$\text{number of crystals } N_{1,2} \text{ in the size range } L_1 \text{ to } L_2 = \int_{L_1}^{L_2} n \, dL \tag{7.3}$$

$$\text{total number of crystals in the distribution, } N_T = \int_{0}^{\infty} n \, dL. \tag{7.4}$$

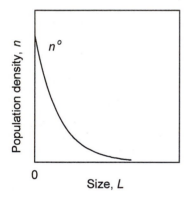

Fig. 7.2 Number (or population) density distribution, also referred to as the differential distribution.

The 'average' size is an important property of the distribution. A number of different averages can be defined. For example, the size at which half the distribution is larger and half smaller is the *median*.

A family of mean sizes can be represented by the general equation

$$\bar{L}_{j,0} = \left[\int_{0}^{\infty} L^j n dL \bigg/ \int_{0}^{\infty} n dL \right]^{1/j} = \left[\mu_j / \mu_0 \right]^{1/j} \tag{7.5}$$

where $\mu_j = \int_{0}^{\infty} L^j n dL$ is the jth moment of the distribution. When $j = 1$ the corresponding length-averaged size $\bar{L}_{1,0}$ represents the size which, when multiplied by the total number of crystals, gives the total length of crystals in the distribution.

Another family of average sizes can be defined through the general definition of the average value of a property:

$$\bar{L}_{j,j-1} = \int_0^\infty L^j n dL \bigg/ \int_0^\infty L^{j-1} n dL = \mu_j/\mu_{j-1}. \qquad (7.6)$$

A commonly used measure of the width of the distribution around the mean size is the coefficient of variation, cv, equal to the standard deviation of the distribution divided by the mean size.

7.3 The continuous mixed suspension mixed product removal crystallizer

Basic equations

In order to develop equations for the size distribution within a crystallizer we will first consider the very simplest type of crystallizer. This is analogous to the continuous stirred tank reactor (CSTR) or back-mix reactor of reaction engineering (Metcalfe, 1997), although in the crystallization literature it has become known as the continuous mixed suspension mixed product removal or MSMPR crystallizer. We make the following assumptions:

(a)   The crystallizer operates continuously.
(b)   Operation is at steady state.
(c)   The crystallizer is perfectly mixed with respect to both the liquid and solid phases.
(d)   The outlet (product) from the crystallizer is identical in all respects to the crystallizer contents.
(e)   There are no crystals in the feed stream to the crystallizer.
(f)   There is no crystal breakage or agglomeration within the crystallizer.
(g)   All crystals have the same shape.
(h)   The linear growth rate of the crystals is defined by $G = dL/dt$ and all crystals grow at the same rate irrespective of their size. This assumption of size-independent growth is sometimes known as McCabe's ΔL law (McCabe, 1929).

The continuous MSMPR crystallizer, which is shown schematically in Fig. 7.3, has a volume V (m³) and is fed with solution at a volumetric flow rate Q (m³ s⁻¹). Production of supersaturation within the crystallizer gives rise to nucleation and subsequent growth; at this stage it is not necessary to specify how supersaturation is achieved. A size distribution of crystals, represented by the population density $n(L)$, develops within the crystallizer. This is of course also the size distribution of crystals in the product stream (see assumption d). Figure 7.4 represents these nucleation and growth processes taking place along the crystal size axis.

The number density can be determined from experimental measurements of the size distribution. For example, if the size distribution has been determined by sieve analysis let us suppose that a mass of crystals w is retained between two adjacent sieves of aperture sizes L_1 and L_2. We take the mean size of this mass as

$$\bar{L} = \tfrac{1}{2}(L_1 + L_2)$$

and define the size range ΔL as

$$\Delta L = L_1 - L_2.$$

The number of crystals in the size range L_1 to L_2 is equal to the mass of all the crystals in mass w divided by the mass of one of the crystals. Defining the volume shape factor of the crystals to be α, the mass of one of the crystals is

$$\alpha \rho_{cry} \bar{L}^3.$$

The number of crystals in the size range L_1 to L_2 is thus

$$w/\alpha \rho_{cry} \bar{L}^3$$

and the number density is given by

$$n = w/\alpha \rho_{cry} \bar{L}^3 \Delta L V$$

where V is the volume of suspension from which the crystal sample of mass w was obtained.

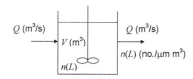

Fig. 7.3 Schematic of continuous MSMPR crystallizer.

The assumption of size-independent growth can be shown to be reasonable using the following argument. Let the overall mass growth rate R_G of an individual crystal of mass m be represented by

$$R_G = \frac{1}{a}\frac{dm}{dt}$$

where a is the surface area of that crystal. We can write

$$a = \beta L^2$$

and

$$m = \alpha \rho_{cry} L^3$$

where α and β are the area and volume shape factors respectively; ρ_{cry} is the crystal density. Thus

$$\frac{dm}{dt} = 3\alpha \rho_{cry} L^2 \frac{dL}{dt}$$

and so

$$R_G = \frac{3\alpha \rho_{cry} L^2}{\beta L^2}\frac{dL}{dt} = \frac{3\rho_{cry}}{F}G$$

where F is the overall shape factor equal to β/α.

The linear growth rate G is thus directly proportional to the mass growth rate R_G. This assumption that G is independent of crystal size is frequently valid.

Note the difference between the linear growth rate of a crystal, G, defined as dL/dt, where L is the overall size of the crystal, and the linear rate of advance of an individual face, v.

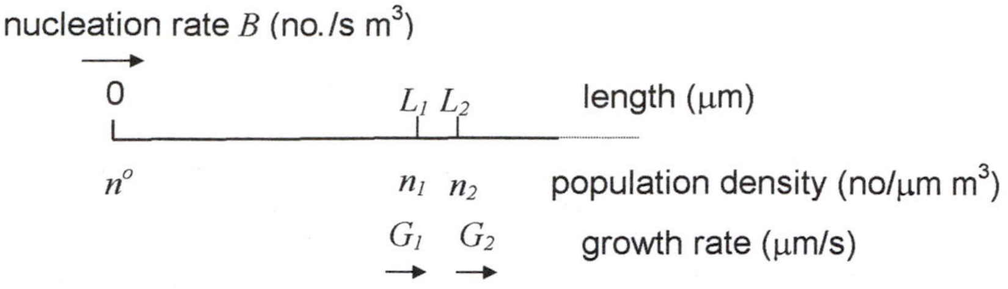

We have seen in Chapter 3 that critical nuclei have sizes of a few tens of molecules, say up to about 10 nm. Compared to typical crystals in a crystallizer which are of the order of 100 μm, these can be considered as being of 'zero size'.

nucleation rate B (no./s m^3)

Fig. 7.4   Nucleation and growth along the crystal size axis.

We now write a number balance over the size range L_1 to L_2. In any time interval Δt, crystals enter and leave this range by growth while some also leave in the product stream. These number flows can be evaluated as follows:

$$\text{number entering by growth} = n_1\, V\, G_1 \Delta t$$

$$\text{number leaving by growth} = n_2\, V\, G_2 \Delta t$$

$$\text{number removed in the product stream} = Q\, \bar{n}\, \Delta L\, \Delta t.$$

A number balance requires that in the steady state (assumption b above) the input numbers equate to the output numbers and so

$$n_1 V G_1\, \Delta t = n_2 V G_2\, \Delta t + Q\bar{n}\, \Delta L\, \Delta t,$$

i.e.

$$\frac{n_2 G_2 - n_1 G_1}{\Delta L} = -\frac{Q}{V}\bar{n}$$

or, in the limit as $\Delta L \to 0$,

$$\frac{d(nG)}{dL} = -\frac{Qn}{V}. \tag{7.7}$$

The mean residence time within the crystallizer is defined by $\tau = V/Q$ and if the growth rate is not a function of crystal size (assumption h) we can write the above equation as

$$\frac{dn}{dt} = -\frac{n}{G\tau}. \tag{7.8}$$

This is the key differential equation describing the steady-state size distribution in a continuous MSMPR crystallizer with all the assumptions listed above. It can be integrated to give the size distribution by using the boundary condition $n = n^o$ at $L = 0$:

$$\ln n = \ln n^o - \frac{L}{G\tau} \quad \text{or} \quad n = n^o \exp\left(-\frac{L}{G\tau}\right). \tag{7.9}$$

The number density at zero size, n^o, can be related to the nucleation rate B by noting that the nucleation rate represents the rate at which new crystal numbers are created at zero size. So

$$B = \left.\frac{dN}{dt}\right|_{L=0} = \left.\frac{dN}{dL}\right|_{L=0} \left.\frac{dL}{dt}\right|_{L=0} = n^o G \tag{7.10}$$

which enables Eqn 7.9 to be written as

$$\ln n = \ln\left(\frac{B}{G}\right) - \frac{L}{G\tau} \quad \text{or} \quad n = \frac{B}{G}\exp\left(-\frac{L}{G\tau}\right). \qquad (7.11)$$

These two alternative forms, Eqns (7.9) and (7.11), describe the steady-state size distribution produced by a continuous MSMPR crystallizer and so define $n(L)$ for such a system. They predict that the number density decays exponentially with size at a rate determined by the growth rate and the mean residence time. There is no empiricism involved in these equations. The size distribution is completely described in terms of physically meaningful parameters—the kinetics of the nucleation (usually secondary nucleation as discussed in Chapter 3) and the growth processes that are taking place in the crystallizer, together with the mean residence time in the vessel. The most convenient way to represent these equations graphically is in the semi-logarithmic form shown in Fig. 7.5.

These equations and the corresponding plots can be used in two alternative ways. If the kinetics are known the size distribution can be predicted. Alternatively, and more usefully at this stage, the equations suggest a method by which the kinetics of nucleation and growth, B and G, can be determined. If the steady-state size distribution is measured in a continuous MSMPR crystallizer and plotted as the number density against size in the semi-logarithmic form of Fig. 7.5, the growth rate can be determined from the slope and the nucleation rate from the intercept at $L = 0$. An example of such data is shown in Fig. 7.6.

Many studies of this sort have been made (e.g. Garside and Shah, 1980). The effect of factors such as solids concentration, stirrer speed, temperature and impurity concentration on the nucleation and growth kinetics can be determined in this way. The technique has the advantage that the kinetics are determined simultaneously in a continuously nucleating environment, conditions similar to those found in industrial crystallizers.

Other distributions

The distribution represented by Eqns (7.9) and (7.11) is that of the number density. The same distribution can be written in other forms by simple algebraic manipulation of these equations, so enabling other properties of these distributions to be calculated. For example, the total number of crystals in the distribution, N_T, can be evaluated by integrating the number of crystals in the size increment dL over all possible sizes, i.e.

$$N_T = \int_0^\infty n\,dL = \int_0^\infty n^o \exp(-L/G\tau)dL = n^o G\tau = B\tau. \qquad (7.12)$$

Similarly, the total length, area and mass in the distribution can be evaluated respectively from

$$L_T = \int_0^\infty nL\,dL = n^o(G\tau)^2 = BG\tau^2 \qquad (7.13)$$

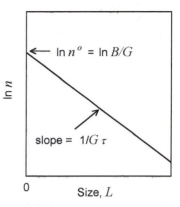

Fig. 7.5 Number density plot.

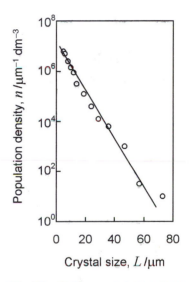

Fig. 7.6 Calcium oxalate is a major constituent of kidney stones. This figure shows the steady-state size distribution represented as a number density plot for precipitation of calcium oxalate in a continuous MSMPR crystallizer (Garside *et al.*, 1982).

$$A_T = \int_0^\infty \beta n L^2 \, dL = 2\beta n^o (G\tau)^3 = 2\beta B G^2 \tau^3 \qquad (7.14)$$

$$M_T = \int_0^\infty \rho_{cry} \alpha n L^3 \, dL = 6\rho_{cry} \alpha n^o (G\tau)^4 = 6\rho_{cry} \alpha B G^3 \tau^4. \qquad (7.15)$$

For practical purposes the mass distribution is the most widely used; after all, crystals are generally sold by mass not number.

The cumulative mass fraction distribution is evaluated by dividing the mass smaller than a given size L by the total mass in the distribution. Representing size by the dimensionless parameter $x = L/G\tau$, the mass fraction distribution $M(x)$ is given by

$$M(x) = M/M_T = \int_0^x x^3 \exp(-x) dx \Big/ \int_0^\infty x^3 \exp(-x) dx$$

$$= 1 - \left(1 + x + \frac{1}{2}x^2 + \frac{1}{6}x^3\right)\exp(-x). \qquad (7.16)$$

By differentiating this equation the differential mass fraction distribution $m(x)$ can be evaluated:

$$m(x) = \frac{dM(x)}{dx} = \frac{1}{6}x^3 \exp(-x). \qquad (7.17)$$

These two forms of the mass distribution are shown in Fig. 7.7.

We can now readily evaluate various mean sizes of this mass distribution. The median size $\bar{L}_M$ is given by equating Eqn (7.16) to 0.5 and solving for x, which gives

$$\bar{x}_M = \frac{\bar{L}_M}{G\tau} = 3.67, \quad \text{i.e.} \quad \bar{L}_M = 3.67 G\tau. \qquad (7.18)$$

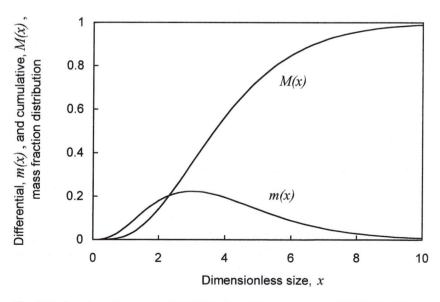

Fig. 7.7   Steady-state mass fraction distributions from a continuous MSMPR crystallizer.

Table 7.1 Average sizes of the crystal size distribution obtained from a continuous MSMPR crystallizer

i	1	2	3	4	2	3	4
j	0	0	0	0	1	2	3
$\bar{L}_{i,j}/G\tau$	1	1.41	1.82	2.21	2	3	4

The mean size $\bar{L}_{3,0}$ (see Eqn 7.5) is equal to the mode or dominant size $\bar{L}_D$. This is the most frequently occurring size and hence corresponds to the maximum in the differential distribution. It is given by

$$\bar{x}_D = \frac{\bar{L}_D}{G\tau} = 3, \quad \text{i.e.} \quad \bar{L}_{3,0} = \bar{L}_D = 3G\tau, \tag{7.19}$$

while the mass average size is given by analogy with Eqn (7.6) as

$$\bar{L}_{4,3} = \left[\int_0^\infty L^4 n dL \Big/ \int_0^\infty L^3 n dL \right] = 4G\tau. \tag{7.20}$$

The coefficient of variation for the mass distribution can be calculated to be 0.5.

The various average sizes from a continuous MSMPR crystallizer are given in Table 7.1.

The form of the mass distributions in Fig. 7.7 is particularly interesting. The reference crystal size $G\tau$ appearing in the dimensionless size $x = L/G\tau$ can be thought of as the size to which a nucleus grows in one residence time. A size of $x = 1$ thus represents a crystal that has been in the crystallizer for just one residence time. But most of the mass is associated with crystals for which x is significantly greater than one, indeed only about 2% of the mass has been present for one residence time or less as can be seen from Fig. 7.7. It was shown above that the mode of the mass distribution is $x = 3$, corresponding to crystals that have been present in the crystallizer for three residence times, while much of the mass corresponds to crystals that have been in the crystallizer for seven or more residence times. This apparent paradox arises because the bulk of the mass within the distribution is concentrated in the larger sizes present whereas most of the numbers are concentrated at the smaller sizes.

The mass balance

Up to now we have only considered the number balance. To provide a full description of the MSMPR crystallizer we must also consider the implications of the mass balance. Figure 7.8 is a schematic that defines the necessary symbols. A mass balance on the solute in solid and solution phases gives

$$Q_{in}c_{in} - Q_{out}c_{out} = Q_{out}M_T.$$

Because the mass deposition arising from nucleation is negligible, the right-hand side of this equation can be taken to represent the rate of growth onto the total crystal surface in the crystallizer:

$$Q_{out}M_T = R_G A_T V = \frac{3\rho_{cry}}{F} G A_T V.$$

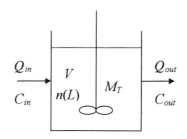

Fig. 7.8 Schematic of continuous MSMPR crystallizer showing mass balance considerations.

With $\quad A_T = \beta \int_0^{\infty} L^2 n dL = 2\beta n^o (G\tau)^3 \quad$ and $\quad \tau = V/Q_{out}$

and assuming $Q_{in} \approx Q_{out}$, these equations can be combined to give

$$G = \frac{F(c_{in} - c_{out})}{3\rho_{cry}\beta\tau \int_0^{\infty} L^2 n dL} = K\frac{(c_{in} - c_{out})}{\tau A_T} \tag{7.21}$$

where K combines the crystal shape factors and crystal density, all constant for a given system. If we assume that the growth rate G is a simple function of the supersaturation Δc then

$$\Delta c = K_1 \frac{(c_{in} - c_{out})}{\tau A_T}. \tag{7.22}$$

K_1 is a combination of K and the constant linking growth rate with supersaturation.

Equation (7.22) demonstrates that the supersaturation at which the crystallizer operates depends on three variables. First, the supersaturation is directly proportional to the difference in solute concentration in the solution entering and leaving the crystallizer. The inlet concentration clearly depends on the upstream processes while the outlet concentration depends on the way in which the crystallizer is operated—on the temperature in the case of a cooling crystallizer, on the boil-up rate with an evaporative system. There is an inverse proportionality with the second variable, the residence time; a longer residence time allows more growth and nucleation to occur and hence leads to a lower solution supersaturation. The third variable, total crystal surface area, also shows an inverse proportionality. Larger values provide more surface area onto which growth can occur and so result in a lower supersaturation.

All this information is summarized in the information flow chart in Fig. 7.9 (Randolph and Larson, 1988). This is essentially the same as that shown in

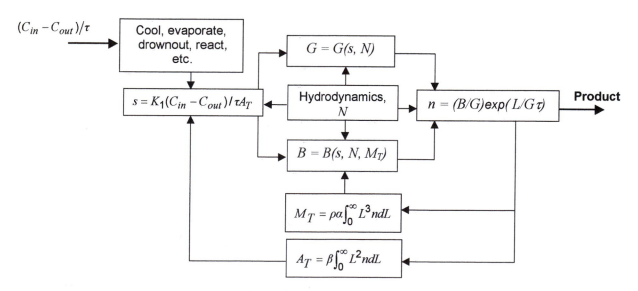

Fig. 7.9   Information flow diagram for an MSMPR crystallizer.

Fig. 1.2 but now we have developed the quantitative relationships between the different variables. The chart highlights the feedback between the CSD, through the corresponding crystal surface area, and the prevailing super-saturation. A second feedback loop arises from the dependence of the secondary nucleation rate on the crystal mass concentration as described in Chapter 3. We will see in Chapter 8 how these feedback loops exert a dominant influence on the behaviour of crystallizers.

References

Bransom S. H., Dunning W. J. and Millard B. (1949) *Discuss. Faraday Soc.* **5** 83.

Garside J. and Shah M. B. (1980) *IEC Proc. Des. Dev.* **10** 509.

Garside J., Brecevic Lj., Mullin J. W. (1982) *J. Crystal Growth* **57** 233.

McCabe W. L. (1929) *IEC* 21 30.

Metcalfe I. S. (1997) *Chemical reaction engineering*, Oxford Chemistry Primer, No 49, Oxford University Press.

Randolph A. D. and Larson M A (1988) *Theory of particulate processes*, 2nd edn, Academic Press, New York.

8 Characteristics of continuous and batch crystallizers

8.1 Continuous MSMPR crystallizers

We have seen in Chapter 7 that the continuous MSMPR crystallizer produces a crystal size distribution that has a fixed shape, one measure of which is a coefficient of variation for the mass distribution of 0.5. The mean size of the distribution can be varied and is always given by a simple multiple of the product $G\tau$. One might instinctively expect that an increase in the residence time τ would result in an increase in the mean crystal size. Examination of the information flow diagram in Fig. 7.9, however, suggests that more subtle effects must be considered.

An increase in the residence time, for example, also results in a decrease in the solution supersaturation within the crystallizer as is specified by Eqn (7.22). This lower supersaturation will result in a lower growth rate (as well as a lower nucleation rate). Thus, although the crystals stay in the crystallizer for longer, they are growing more slowing during this longer time. The net result will be either an increase or a decrease in the mean crystal size, the outcome depending on whether or not the increase in residence time outweighs the decrease in growth rate.

It is possible to determine the overall effect of changes in residence time on the average crystal size in a quantitative way by combining some of the previously derived equations.

We have seen that secondary nucleation kinetics can be represented by an equation of the form

$$B = k_b M_T \Delta c^b$$

while the growth rate can be written

$$G = k_g \Delta c^g.$$

It is often convenient to eliminate the concentration driving force between these two equations and write the nucleation rate as a function of the growth rate:

$$B = K_R M_T G^i$$

where the relative rate constant

$$K_R = k_b / k_g^i$$

and the relative kinetic order i is given by

$$i = b/g.$$

Let us use the mode of the mass distribution $\bar{L}_D$ to characterize the mean crystal size and recall that $\bar{L}_D = 3G\tau$ (Eqn 7.19). The production rate of the crystallizer is given by the product of the volumetric flow rate through the vessel, Q, and the total concentration of crystals, M_T, in the slurry; M_T is given by Eqn (7.15):

$$M_T = 6\rho_{cry}\alpha BG^3\tau^4.$$

If the relative crystallization kinetics are represented by

$$B = K_R M_T G^i,$$

rearrangement of these equations shows that

$$\tau = \frac{\bar{L}_D}{3}\left[\frac{2\rho_{cry}\alpha K_R \bar{L}_D{}^4}{27}\right]^{1/(i-1)}. \tag{8.1}$$

Since the crystallizer volume $V = Q\tau$, this equation in effect gives the crystallizer volume required to produce a size distribution having the mean size $\bar{L}_D$. The equation also shows that the mean size and residence time are related by

$$\tau \propto \bar{L}_D{}^{(i+3)/(i-1)}. \tag{8.2}$$

This relationship is illustrated in Fig. 8.1 in which the subscripts 1 and 2 correspond to the mean sizes obtained at two corresponding residence times. If $i = 1$ it can be shown that:

$$\bar{L}_D = \left[\frac{27}{2\rho_{cry}\alpha K_R}\right]^{1/4} \tag{8.3}$$

and the mean size is independent of residence time. In this case the increased time that the crystals spend in the crystallizer is exactly offset by the lower growth rate that they experience during their residence in the vessel. Under these circumstances Eqn (8.3) indicates that the only way to change the crystal size is by changing the relative rate constant K_R.

8.2 Size-dependent removal rates

The 'mixed product removal' assumption of the MSMPR crystallizer ensures that crystals of all sizes have the same mean residence time, which is also equal to that of the solution. If crystals have mean residence times that depend on their size, considerable extra flexibility of operation is achieved and greater variations in the CSD are possible.

Fines removal

We have seen that increasing the residence time in an MSMPR crystallizer is an ineffective way of increasing crystal size. Far more effective in addressing this objective is the removal of a large proportion of the *numbers* of the small, or fine, crystals before they have grown to a size where they contribute significantly to the *mass* of the crystal product. The net effect of this strategy is to force the supersaturation, and hence the growth rate, to higher levels, so producing the same production rate on fewer crystals of larger average size.

The essential features of a fines removal system can be illustrated by the simple model shown in Fig. 8.2. Size classification results in crystals smaller than the 'cut' size L_F being removed preferentially in a stream with flow rate Q_F. This is often achieved by the presence of a baffle that forms a calming section. Here the larger crystals settle out while the upflowing liquor containing fine crystals is removed. The largest crystal size removed in this stream is determined by the solution velocity and the terminal settling velocity of the crystals but typically may be between 50 and 100 μm. Crystal product comprising all crystal sizes is removed at rate Q_P. The mean residence times of the fines τ_F and of the product τ_P are then:

$$\tau_F = \frac{v}{Q_F + Q_P} \quad (L \leq L_F); \quad \tau_P = \frac{V}{Q_P} \quad (L > L_F). \tag{8.4}$$

The population density distribution of the fine crystals is clearly given by

$$n = n^o \exp(-L/G\tau_F) \quad \text{for } L \leq L_F \tag{8.5}$$

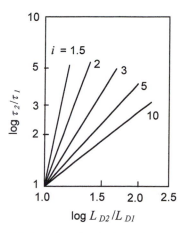

Fig. 8.1 Relation between the residence time and mean size for an MSMPR crystallizer. Note the great insensitivity of size to changes in residence time for low values of the relative growth order i. For example if, $i = 2$ and an increase in mean size of 20% is required (i.e. $L_{D2}/L_{D1} = 1.2$), the required ratio of residence times would be 2.5. This would require a reduction in production rate of the same ratio, a penalty that is unlikely to be economic. The mean crystal size produced in an MSMPR crystallizer for a given system is therefore determined principally by the kinetics (i.e. by the values of K_R and i) rather than by the residence time.

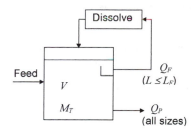

Fig. 8.2 Fines removal system. Note that the fines are dissolved and returned to the crystallizer.

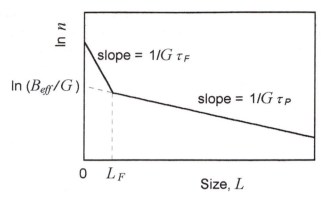

Fig. 8.3 Population density plot for an idealized fines removal system. The population density of fines decays at a rate that depends on the fines residence time τ_F.

while if we assume that $L_F \ll \bar{L}_D$, the population density distribution for the rest of the size range is

$$n = n^o \exp(-L_F/G\tau_F) \exp(-L/G\tau_P) \quad \text{for } L > L_F. \tag{8.6}$$

The overall size distribution represented by these two equations is shown in Fig. 8.3.

We can think of the constant term in Eqn (8.6) as an effective nuclei population density and so the effective nucleation rate is:

$$B_{eff} = Gn^o \exp(-L_F/G\tau_F). \tag{8.7}$$

Since the *mass* of crystals smaller than L_F is negligible, the mass distribution of the product can be calculated by considering only that part of the distribution larger than L_F. Equation (8.7) gives the 'effective' nucleation rate for the distribution.

Using an analogous derivation to that for Eqn (8.1) we can now show that the product residence time required to produce an average size $\bar{L}_D$ when fines removal is in place is:

$$\tau = \frac{f\bar{L}_D}{3}\left[\frac{2\rho_{cry}\alpha K_R\bar{L}_D^4}{27}\right]^{1/(i-1)} \tag{8.8}$$

where

$$f = \left[\exp(-3L_F\tau_P/\bar{L}_D\tau_F)\right]^{1/(i-1)}. \tag{8.9}$$

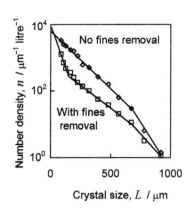

Fig. 8.4 The experimental data shown here were obtained from a crystallizer with a fines removal system. The form of the data is very similar to the idealized model. This figure shows results for crystallization of potassium nitrate from aqueous solution (after Juzaszek and Larson, 1977).

Typical values for a fines removal system might be $\tau_P/\tau_F = 10$, $L_F/\bar{L}_D = 0.1$ and $i = 3$, so giving $f = 0.22$. The residence time required to produce a given product average size is thus reduced to less than one-quarter of the value for an MSMPR crystallizer. Alternatively, for a given residence time the average product size achieved with a fines removal system is greater than in the standard MSMPR configuration.

Inevitably, a number of disadvantages arise when fines removal systems are employed. The higher operating supersaturation can lead to greater encrustation on the walls of the crystallizer, there is a higher possibility of instability, and the economics are often poor since dissolution of the fines by either heating or dilution must be accomplished. Nevertheless, the greater operating flexibility makes this mode of operation very attractive and it is widely employed industrially.

Classified product removal

The bulk of the product mass occurs in the larger size ranges of the CSD. If these larger sizes have residence times that are a function of their size, the resulting 'classified product removal' can have a profound effect on the operation of the crystallizer. The larger crystals may have a shorter or longer mean residence time than the smaller crystals and so the slope of the population density plot in the classified region will be higher or lower, respectively, than that in the smaller size range (Fig. 8.5).

A shorter mean residence time for the larger crystals may arise if they classify preferentially into the region of the crystallizer from which the product is removed. This may be deliberately designed for, using an elutriation leg for example, or it might occur inadvertently through poor control of local solution velocities. In either case the product size distribution is narrowed compared to the corresponding MSMPR case. The mean crystal size, however, is usually smaller. This result might appear surprising but it occurs because, for a given production rate, the product now contains a smaller *number* of crystals. In an MSMPR crystallizer large numbers of fine crystals are removed in the product so that a higher nucleation rate can be tolerated. The smaller number of crystals in the classified product can only be sustained if the nucleation rate is lower. This demands a lower supersaturation and hence lower growth rate. In practice, product classification is often combined with fines removal, the latter removing the excess fines that result from operating at the higher supersaturation necessary to maintain the required product size.

Longer mean residence times for the larger crystals usually result when the product off-take is some distance from a region into which larger crystals classify. Product CSDs are then wider than in the MSMPR case. Such a mode of operation frequently leads to unstable conditions; the larger crystals gradually accumulate in the classifying region until a large proportion of them is rapidly removed in the product. The resulting rapid decrease in crystal surface area leads to a corresponding increase in supersaturation and hence in the nucleation rate.

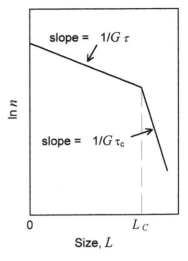

Fig. 8.5 The shape of the population density distribution for classified product removal follows directly from the mean residence time in the larger size ranges. In this case crystals larger than L_C have a shorter mean residence time, τ_C, than that of the smaller crystals, τ, and so the number of crystals larger than L_C decays more rapidly than those in the smaller size range.

8.3 Other factors influencing crystal size distribution

Size-dependent growth rate

Up to this point we have assumed that crystals of all sizes have the same growth rate. Conditions can exist where this is not the case. For example, if the mass transfer process controls the growth rate (see Section 4.4), the larger crystals would be expected to grow more rapidly than the smaller because of the enhanced mass transfer coefficient that results from the higher relative velocities between large crystals and the solution. An example of this is shown in Fig. 4.15. There is also evidence that very small crystals in the range below about 10 μm often have lower average growth rates than the larger sizes. This is possibly because many of them are bounded by faces which do not possess dislocations(see Fig. 4.9), or because they are so strained that they have solubilities that are higher than that of larger crystals (Ristic *et al.*, 1990).

The slope of the population density plot again directly reflects the variation of growth rate with size. If the growth rate increases with crystal size, the slope of the population density plot ($-1/G\tau$) decreases with size. The plot is now no longer linear and the curvature follows the growth rate variation.

Agglomeration and breakage

Agglomeration is a frequent occurrence in many crystallizing systems, particularly within the smaller size range. The net result is that the distribution contains more crystals of larger sizes, and fewer of smaller sizes, than would be found in the normal MSMPR distribution. A non-linear population density plot thus results, the shape of which is similar to that obtained for size-dependent growth when growth rate increases with size.

Crystal breakage can be thought of as the reverse of agglomeration. Now there is a deficit of large crystals because some of them have been broken, the breakage fragments populating the smaller sizes and so increasing the numbers of small crystals. The population density plot is again non-linear but now with the opposite curvature, the negative slope increasing with increasing size.

Incorporation of agglomeration or breakage into the population balance requires the use of birth and death functions. A full mathematical analysis of either of these cases is rather difficult since the particle growth process is now discontinuous with size, unlike the continuous process of normal crystal growth. The size distribution of those crystals contributing to an agglomeration event or resulting from a breakage event must be known and the equations to be solved must include the requirement that the total crystal volume is conserved during the agglomeration or breakage event. Illustrative examples of work in this area are given by Randolph and Larson (1988) for breakage and by David *et al.* (1991) for agglomeration.

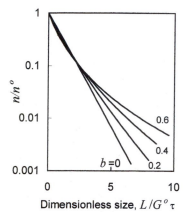

Fig. 8.6 An empirical equation that has proved useful in describing size-dependent growth has been suggested by Abegg *et al.* (1968):

$$G = G_0(1 + \gamma L)^b \quad \text{for } b < 1$$

where G_0, b and γ are constants. Corresponding population density plots can be calculated and are shown here for various values of b. When $b = 0$ the growth rate becomes independent of size and the exponential decay of the MSMPR equation holds. For $b > 0$ the growth rate increases with crystal size.

8.4   Dynamics and stability of continuous crystallizers

A description of the unsteady-state behaviour of a continuous crystallizer requires solution of the unsteady-state population balance equation. Retaining all the assumptions made in deriving the MSMPR equation shown in Eqn 7.8 except that of steady state gives the appropriate equation:

$$\frac{\partial n}{\partial t} + G\frac{\partial n}{\partial L} = -\frac{n}{\tau}. \qquad (8.10)$$

Transients in the CSD are caused by disturbances that have their origin outside the crystallizer; examples are changes in feed rate, feed composition, feed temperature and boil-up rate. The course of such transients can be calculated from Eqn (8.10) and demonstrates how long it takes for the crystallizer to regain a new steady state. This time has been shown to be typically between 8 and 10 residence times (see, for example, Randolph and Larson, 1962). Such a long period is a direct result of the feedback between changes in supersaturation and crystal area illustrated in Fig. 7.9. An example of calculated transient behaviour is shown in Fig. 8.8.

Instabilities are linked to the interactions between the inherent kinetics of the crystallizing system and the residence time distribution function. It is possible to envisage that the feedback loop to supersaturation, through the crystal surface area (Fig. 7.9), could result in changes to the supersaturation, and hence to other variables, of ever-increasing magnitude; such a system would be inherently unstable. The classic paper of Randolph and Larson (1962) showed that such a situation would only occur in an MSMPR crystallizer if the relative kinetic order $i > 21$. Such a high value is extremely unlikely and so MSMPR crystallizers are generally inherently stable and will eventually return to a steady state following a disturbance. Instability could occur, however, if there are discontinuous changes in the nucleation rate as may happen, for example, if changes in supersaturation cause the nucleation mechanism to change from secondary to primary.

Many industrial crystallizers exhibit low-frequency cycling that persists over long periods of time. This is usually associated with the use of product classification and is exacerbated if this is combined with a fines removal system. A number of such examples are discussed in the literature (for example Randolph *et al.*, 1977; Eek *et al.*, 1995).

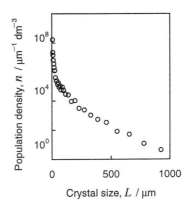

Fig. 8.7 When the population density distributions in MSMPR crystallizers are measured down to sizes below about 10 μm, curvature is frequently observed in the plot of ln(*n*) against *L*. It is thought that this most usually arises not from size-dependent growth but from 'growth rate dispersion' in which the growth rate of crystals of any one size varies about some average value. Given the structure-sensitive nature of the growth process discussed in Chapter 4, this is not surprising. Consequently, a large fraction of the crystals at very small sizes has an extremely low growth rate.

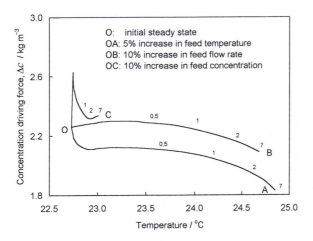

Fig. 8.8 Simulation of the transient behaviour of a continuous MSMPR crystallizer. This diagram illustrates the calculated trajectories followed within the crystallizer in the supersaturation–temperature plane following step changes in three different variables, feed flow rate, feed temperature and feed concentration. The numbers on each curve represent the number of residence times required to arrive at the specific point after the step change. (After Akoglu *et al.*, 1984.)

8.5 Batch crystallizers

Batch crystallizers are very widely used, particularly in the production of speciality, agro- and pharmaceutical chemicals where the production rates are comparatively low. They are simple, flexible and require less capital investment than continuous units. Whereas continuous crystallizers are usually purpose-built for a specific application, batch crystallization operations tend to be carried out in a multipurpose plant, the same vessel serving as a reactor, crystallizer, mixer and blender, as well as handling many different products. In discussing batch crystallization, the emphasis thus tends to be on the mode of operation rather than on the design of a specific piece of equipment.

A wide range of methods is used to produce supersaturation in batch processing. Cooling and evaporation are frequently encountered, but with organic materials, in particular, drowning out crystallization is very common (see Chapter 2). This involves adding a miscible non-solvent to reduce the solubility of the desired crystalline product in its mother liquor. This third component may be water if the product is dissolved in a polar organic solvent, a polar organic solvent if the product is in aqueous solution (see for example Fig. 2.4), an acid or base if the solubility is pH dependent resulting, for example, from a zwitterionic form of the product, or a salt to make use of the common ion effect (salting out). Reaction crystallization, where the reaction product is relatively insoluble and crystallizes directly following a chemical reaction, is also frequently encountered.

All batch crystallizations take place under unsteady-state conditions. Central to their successful operation is the control of supersaturation during the batch. This depends on the operating policy which specifies the batch time, the rate at which supersaturation is generated, the possible addition of seed crystals, and the type and level of agitation.

The mass balance

Consideration of a simple mass balance defines the factors that control crystallization rate. A mass balance on the solute can be written

$$\frac{dM}{dt} + \frac{d(Vc)}{dt} = \frac{dM}{dt} + V\frac{dc}{dt} + c\frac{dV}{dt} = 0 \tag{8.11}$$

where M is the amount of crystal in suspension, V is the solution volume and c is the concentration of solute in solution.

In a cooling crystallization the solution volume is approximately constant and we will assume that the solution concentration is always close to the saturation value c_{eq}. For this case the above equation can thus be written

$$\frac{dM}{dt} + V\left(\frac{dc_{eq}}{d\theta}\frac{d\theta}{dt}\right) = 0. \tag{8.12}$$

The first derivative in the bracket represents the slope of the solubility curve and is a characteristic of the system being crystallized, while the second term is the cooling rate. It is therefore the cooling rate that determines the rate of crystallization, dM/dt.

In evaporative crystallization the solution concentration remains approximately constant at a value close to the equilibrium point at the evaporation temperature. Equation (8.11) can therefore be written

$$\frac{dM}{dt} + c_{eq}\frac{dV}{dt} = 0. \qquad (8.13)$$

In this case the crystallization rate is determined by the evaporation rate dV/dt.

These two examples demonstrate the importance of the rate of production of supersaturation, as determined by the rate of cooling or of evaporation, in controlling the crystallization. This rate is the key control variable in batch crystallization. The same principle can be extended to crystallization produced by any other means, by drowning out crystallization for example, where the rate of addition of the third component is the key control variable.

Supersaturation control and seeding

We will take cooling crystallization as an example and recall the solubility/supersolubility diagram and the concept of the metastable zone (Fig. 8.9). Rapid cooling will inevitably cause the solution to cross the boundary of the metastable zone, resulting in uncontrolled nucleation. The very large numbers of nuclei so produced set a limit on the size to which they can grow, since the available solute resources must be distributed over this large number of particles.

Supersaturation control entails ensuring that uncontrolled nucleation does not occur but rather that the supersaturation remains constant, and within the metastable zone, during the whole of the crystallization. Such a process requires a concentration–temperature profile of the form shown in Fig. 8.9.

Even better control of supersaturation and hence of crystal size can often be achieved if seed crystals are added to the solution. Deposition of solute by growth on these seeds rather than by the production of new nuclei is thus

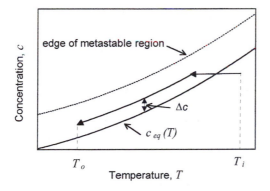

Fig. 8.9 Maintaining a constant supersaturation during the course of a batch crystallization usually enables the nucleation rate to be minimized and maximum growth to take place on the crystals. The result is a larger crystal size together with a narrower size distribution. This effect is further enhanced if the system is seeded just after supersaturation is initially produced. In cooling crystallization the required concentration–temperature profile is shown above.

enhanced. An initial estimate of the amount of seed to add can be made by assuming that all the depletion of supersaturation takes place by growth on the seeds, and that the seed are monodisperse of size L_s. The mass of seeds M_s is then related to the product yield M_p and size L_p by

$$L_p/L_s = \left(M_p/M_s\right)^{1/3}. \tag{8.14}$$

Cooling crystallization

If a crystallizing batch is cooled naturally by, for example, allowing constant temperature cooling water to flow at a constant rate through the cooling jacket of the crystallizer, the temperature–time profile $\theta(t)$ can be calculated as

$$(\theta - \theta_o)/(\theta_i - \theta_o) = \exp(-t/\tau_B) \tag{8.15}$$

where θ_i is the initial temperature and θ_o is the final and hence the cooling water temperature. The batch time constant τ_B is given by

$$\tau_B = M_s C_p/UA_B. \tag{8.16}$$

U is the overall heat transfer coefficient, A_B the heat transfer area and M_s the mass of solution that has specific heat C_p.

A cooling curve of this type gives very high cooling rates initially, when the temperature driving force is greatest. This is exactly what is *not* required to maintain the supersaturation at a constant value within the metastable zone. The rapid initial cooling will produce uncontrolled crystallization and result in the solution inevitably crossing the metastability limit, so producing massive nucleation. Controlled crystallization, on the other hand, demands very low rates of supersaturation production, and hence very low cooling rates in the early stages of the batch, with the cooling rate gradually increasing with time. This ensures that the rate of production of supersaturation always matches the crystal surface area on which the supersaturation is deposited.

Much work has been done to calculate the temperature–time profile required to achieve such controlled crystallization. Assuming an unseeded system and a linear temperature–solubility relation, the cooling curve that must be followed to ensure constant supersaturation during the whole of the batch process was first calculated by Mullin and Nyvlt (1971) to be

$$(\theta_i - \theta)/(\theta_i - \theta_o) = (t/t_B)^4 \tag{8.17}$$

where t_B is the batch time. If the batch is seeded the exponent on (t/t_B) becomes 3 (Nyvlt, 1991). Such controlled cooling curves are almost mirror images of the natural cooling curve (Eqn 8.15), giving the required very low initial cooling rate that then gradually increases as the batch proceeds (Fig. 8.10). Experimental results confirm the advantages to be gained through the increase in product crystal size and the narrowing of the size distribution by following a controlled crystallization.

The occurrence of secondary nucleation sets some limits on the size distribution that can be achieved with controlled crystallization. As soon as crystals appear in the solution secondary nucleation is likely (see Section 3.5) and the number of nuclei that are produced by this mechanism will increase

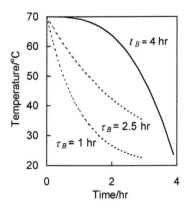

Fig. 8.10 Natural cooling curves give very high initial cooling rates. This is the opposite of that required for controlled crystallization (cf. Fig. 8.9). The controlled cooling curve ensures the necessary very slow rates of supersaturation production during the initial stages of the batch and so minimizes nucleation.

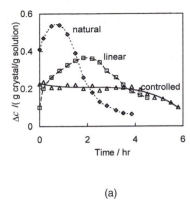

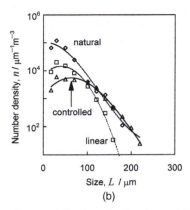

(a)   (b)

Fig. 8.11 Experimental data for the batch cooling crystallization of potassium sulphate demonstrate the advantage of controlled crystallization. Here the results of using three different cooling profiles are shown—natural cooling (Eqn 8.15), a linear cooling profile and controlled cooling (Eqn 8.17). Both natural and linear cooling result in sharp peaks in the concentration driving force (a), whereas the controlled cooling curve produces an almost constant level of supersaturation throughout the batch. The improvement in the product size distribution (b) is evident.

with increasing mass of crystals in the solution. The longer the batch time, therefore, the greater the number of secondary nuclei and the more their presence will widen the size distribution by the development of a population of crystals in the smaller size range. A bimodal size distribution frequently results. If a large product size and narrow distribution are required it is important to minimize secondary nucleation and the most effective way of doing this is usually to minimize the stirrer power input.

Other crystallization methods

The principles that have been outlined for cooling crystallization can be applied to any other method of batch crystallization. In all cases the key is control of the supersaturation throughout the batch by manipulating the rate at which supersaturation is produced. So, in evaporative crystallization, the evaporation rate should be very small at the start of the batch, gradually increasing as a greater crystal surface area develops. In drowning out crystallization the diluent should initially be added very slowly and gradually increased during the course of the batch.

By making a number of simplifying assumptions the required equations for these different time profiles can be calculated (e.g. Tavare *et al.*, 1980). In practice, it is not usually important that the exact form of the theoretical equation is followed, but rather that a reasonable approximation to the appropriate time profile is achieved. Final optimization of the process is invariably achieved through subsequent experimentation.

The population balance for batch crystallization

In principle, it is possible to calculate the full crystal size distribution of the product from a batch crystallization by solving the appropriate population

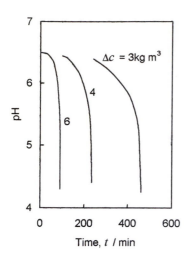

Fig. 8.12 In cases where the pH is changed to achieve changes in solute solubility, controlled crystallization requires that the pH changes very slowly during the early part of the batch (after Zhu and Garside, 1997).

balance equation. For a constant volume batch process in which the contents of the batch are fully mixed this equation is

$$\frac{\partial n}{\partial t} + \frac{\partial (nG)}{\partial L} = 0. \tag{8.18}$$

Much work has been published in this area (see, for example, Tavare, 1987) but it has proved to be of somewhat limited practical value. The partial differential equation (8.18) is not easy to solve and even more difficult is accurate specification of realistic initial and boundary conditions. Appropriate kinetic data, without which system-specific results cannot be obtained, are rarely available. Development of a useful description of the batch process has therefore generally focused on techniques to control the supersaturation as described above.

Practical operating factors in batch crystallization

The need to minimize secondary nucleation has already been alluded to. Nienow (1976) has shown that this is best achieved by using a large slow-moving stirrer rather than a small high-speed device. Whatever the stirrer type employed, it is crucial that the crystals are kept in suspension; crystals that settle to the base of the vessel are no longer exposed to fresh supersaturated solution and tend to agglomerate. Many equations have been suggested for calculating the stirrer speed, N_{js}, required to just suspend crystals of a specific size, L. That suggested by Zweitering (1958) is probably the most satisfactory:

$$N_{js} = S\left[v^{0.1}L^{0.2}X^{0.15}D^{-0.85}(g\Delta\rho/\rho_l)^{0.45}\right] \tag{8.19}$$

where v is the kinematic viscosity of the solution, X is the fraction of solids in the vessel, D is the stirrer diameter, g is the gravitational acceleration, ρ_l is the liquid density and $\Delta\rho$ the difference in density between crystal and liquid. The value of the dimensionless constant S depends on the vessel geometry and values are given in a number of texts (e.g. Mullin, 1992).

References

Abegg C. F., Stevens J. D. and Larson M. A. (1968) *AIChE J.* **14** 118.
Akoglu K., Tavare N. S. and Garside J. (1984) *Chem. Eng. Commun.* **29** 353.
David R., Marchal P., Klein J. P. and Villermaux J, (1991) *Chem. Eng. Sci.* **46** 205.
Eek R. A., Dijkstra Sj., van Rosmalen G. M. (1995) *AIChE J.* **41(3)** 1.
Juzaszek P. and Larson M. A. (1977) *AIChE J.* **23** 460.
Mullin J. W. (1992) *Crystallization*, 3rd edn, Butterworth-Heinemann, Oxford,
Mullin J. W. and Nyvlt J. (1971) *Chem. Eng. Sci.* **26** 369.
Nienow A. W. (1976) *Trans. IChemE.* **54** 205.
Nyvlt J. (1991) in *Advances in industrial crystallization*, J. Garside *et al.*, (eds), p. 197, Butterworth-Heinemann, Oxford.
Randolph A. D., Bechman J. R. and Kraljevich Z. I. (1977) *AIChE J.* **23** 500.
Randolph A. D. and Larson M. A. (1962) *AIChE J.* **8** 639.

Randolph A. D. and Larson M. A. (1988) *Theory of particulate processes*, 2nd edn, Academic Press, New York.

Ristic R. I., Sherwood J. N. and Shripathi T. (1990) *J. Cryst. Growth* **102** 245.

Tavare N. S. (1987) *Chem. Eng. Commun.* **61** 259.

Tavare N. S., Garside J. and Chivate M. R. (1980) *Ind. Eng. Chem. Proc. Des. Dev.* **19** 653.

Zhu J. and Garside J. (1997) *IChemE Jubilee Research Event*, p. 449, IChemE, Rugby.

Zweitering T. N. (1958) *Chem. Eng. Sci.* **8** 244.

9 Crystals in formulated products

Having developed a process for preparing a product in a crystalline form, it is important to appreciate that these crystals usually then have to be transformed into a product. In some cases this is achieved by further chemical reaction, in others by formulation. Formulation is a generic term given to the collection of physical processes such as milling, mixing, dissolution, tableting, and melting that transform individual single components into multicomponent products.

In the commodity chemical industry, crystalline products are commonly manufactured at production rates of millions of tonnes per annum. Terephthalic acid, for example, is specified by its purity and particle size since it is used as a solid reagent to form polyester fibre. Ammonium nitrate is specified by its bulk density and mechanical properties as it is sold for direct application as a fertilizer. Speciality chemicals, on the other hand, rely heavily on formulation. Aspirin tablets contain a fixed amount of the active ingredient acetylsalicylic acid, together will an inert diluent, a binder and a disintegrating agent, the whole mixture being sugar coated. This is thus a complex multicomponent system, sold for its effect (as a painkiller) rather than on the basis of its chemical composition.

9.1 Crystals as active products

Perhaps the most important component of a formulated product is the active ingredient which, although isolated in the production plant as a crystalline material with appropriate characteristics for purity and bulk handling, must now be present in a form suitable for maximizing its effectiveness in its end use. For example, many pharmaceutical materials are formulated as tablets for oral dosage. These tablets are designed to disintegrate in the stomach, releasing crystalline particles of the active drug that then enters the blood stream by dissolution in the gastric fluids and absorption across the gastrointestinal membrane. The rate of dissolution in such a situation is given by

$$\frac{dm}{dt} = \frac{DA}{h}\left(c_{eq} - c\right) \tag{9.1}$$

in which dm/dt is the rate of dissolution, D the diffusion coefficient of the drug molecule, A the surface area of drug crystals, h the thickness of the diffusional layer around each crystal, and c_{eq} and c the saturation and actual concentrations of the drug in the gastric fluids. The form of this equation suggests that the key step in the formulation process will be the reduction of crystal size in order to maximize the surface area of the active drug in the tablet and hence maximize the dissolution rate. Size reduction is usually achieved by milling and grinding processes which impose mechanical forces on the crystals, either using a ball mill or as a result of self-impact in an attritor mill. Such processes typically reduce sizes from 200 to 0.5 μm.

In other situations, for example agrochemicals, molecules with biocidal activity, and transdermal implants of drugs, it is important that the release of the active component from the formulation be controlled so that it occurs over an extended period. This can be done by formulating large crystals or by encapsulating the active crystalline material with a barrier coating which slows its dissolution and diffusion during use.

9.2 Crystals as property controllers

In some formulated products the active component plays a dual role in contributing both to the efficacy and to the overall physical properties of the product. In margarine and chocolate, for example, triglyceride crystals are not only important contributors to the 'mouth feel' of the product but also the size and shape of these crystals contribute to its rheological and mechanical properties.

In other cases the active material is formulated with other ingredients or excipients which, when crystallized, may form an interconnected network encapsulating the active. It is this network that determines the overall mechanical and chemical properties of the product. Catalysts are good examples. A catalyst for a gas-phase reaction must satisfy a number of key requirements. Since it will be charged into a large-scale reactor it must be in a physical form that will not degrade during its lifetime otherwise discharge of spent material becomes problematic. It must have sufficient mechanical robustness to withstand high temperatures and pressures without degrading. Reactants must have easy and sustained access to an active catalytic surface. To satisfy these criteria active catalyst phases (typically crystalline metals or oxides) are mixed with inert support materials (often alumina or silica) and pelleted by compression to yield centimetre-sized pellets with appropriate mechanical and porosity properties. The relative crystal sizes and method of mixing of active and inert phases, together with the pelleting conditions, all have to be optimized to create an acceptable product.

Other examples are found in bars of soap, shoe polish, greases and deodorant sticks where an organic material such as an alkane or long-chain alcohol forms a crystal network to provide the product with required mechanical properties. The active biocides, perfumes or pigments can then be held within this matrix of small crystals and released by mechanical action, rubbing for example, when needed.

The key to achieving an acceptable product in all these cases is in the control of crystallization during product manufacture. In some cases specialist equipment is used, a good example being scraped surface heat exchangers that are employed in both ice cream manufacture and the refining of vegetable oils (composed mainly of triglycerides).

Manufacture of ice cream involves passing a mixture of water, milk, sugars, cream, stabilizers, emulsifiers, flavouring and air through such a heat exchanger. The exchanger has the form of a cylindrical barrel through which the mixture is transported by a revolving helical screw. Cooling at the outer barrel wall initiates nucleation and subsequent growth of ice crystals. The characteristics and acceptability of the ice cream depend in part on the size and size distribution of these crystals; if the size is greater than about 60 μm the 'mouth feel' of the product becomes excessively gritty. It is possible to model the exchanger as a continuous crystallizer but, as we have seen in Chapters 7 and 8, this requires knowledge of the appropriate nucleation and growth kinetics. Knowledge of the mechanism by which the crystals nucleate, the site(s) in the exchanger where nucleation occurs, and an understanding of the path that crystals take on their subsequent passage through the exchanger, are all essential to the development of such a model and hence to control of the ultimate size distribution and achievement of a satisfactory product.

9.3 Crystals causing problems

Formulated products often deteriorate with time due to the growth of crystals. Ice cream continually put into and taken out of the freezer is subject to temperature cycling which causes subsequent melting and solidification of ice. Over a long period of time this leads to coarsening of the ice and loss of creamy texture. Such processes are often referred to as 'ripening' but should not be confused with 'Ostwald ripening', which drives changes of size on the nanometre scale (Eqn 3.10).

With formulated products in which the active is in the form of a crystalline solid there is potential for things to go wrong. In aqueous dispersions, creams and aerosols where micrometre-sized crystals are involved, particle enlargement due to crystal growth commonly destabilizes the product. In a dispersion this would result in settling out of the active to the bottom of its container, in a cream an undesirable grittiness would occur, while in an aerosol the valve would become blocked with the result that the canister becomes inoperable.

Crystal growth in these situations is often driven by temperature cycling during storage with warm days and cold nights causing the active to dissolve in the day and crystallize at night. Over a period of time the smaller crystals in the population disappear while the larger ones grow. As the surface area of crystals decreases, so the dissolution cycle in this process becomes less effective in returning the crystals to their original size and the product changes irreversibly.

Sometimes a polymorphic transformation can occur in the active with the structure as formulated converting to a more stable crystalline polymorph. As discussed in Chapter 6, if the two structures are enantiotropically related, this switch will occur reversibly with cycling temperature. Ammonium nitrate used in explosives and fertilizers provides a good example. The transition temperature between forms IV and III is 32°C and temperature cycling causes the product to fracture into a fine powder which subsequently cakes and can give severe handling problems. For materials in which the structures are related monotropically, the change is irreversible. If the transformation mechanism is through dissolution and crystal growth, an increase in crystal size will occur. The overall result of these processes is that the product will need to be reformulated, perhaps with a growth inhibitor incorporated and certainly from the most stable crystal structure.

9.4 Crystallization controllers as products

This final category refers to molecules that have been developed specifically as products to prevent unwanted crystallization.

One example is found in fuel additives. Middle distillate fuels for use in diesel engines and as heating oil are complex mixtures of hydrocarbon molecules, between 15 and 30% of which are straight-chain paraffins (*n*-alkanes) with chain lengths between about C_9 and C_{28}. When these distillates are subject to low temperatures the heavier paraffins become insoluble and crystallize. The crystals are typically thin plate-like rhombs which can grow as large as 1 mm across. Having formed, such crystals (see Fig. 5.1) rapidly block the in-line fuel filters present in both diesel engines and heating systems, causing fuel starvation and failure of the vehicle or heating system. In addition, attractive colloidal forces between the surfaces of the crystals can lead to the formation of a crystal network. This causes the viscosity of the fuel to rise until, with as little as 1% of paraffins in the crystalline state, the fuel becomes gel-like and can no longer be pumped through the narrow pipelines in diesel engines or heating systems. To prevent these phenomena crystal growth inhibitors and nucleating aids are added to the fuel.

In order to develop such products it is important to have a test against which the activity of selected molecules can be compared. In this way it is possible to develop *structure–activity* relationships for potential product molecules. One such measure in the paraffin application is the *cloud point*, the temperature at which the first wax crystals begin to appear on cooling. A second is the *pour point*, the temperature at which the fuel becomes sufficiently gelled that it no longer behaves as a liquid. A typical additive would be a copolymer of ethylene and vinylacetate which has sequences of alkyl chains separated by more polar groups. The former bind to the crystal while the latter offer a good steric barrier to further crystal growth. The activity of such polymers as a function of their molecular structure can be explored by measuring the amount needed to inhibit crystal growth. Typical data are shown in Fig. 9.1 for the crystallization of n-$C_{32}H_{66}$ in the presence of a series of polymers. These results reveal the importance of alkyl chain length in determining the efficacy of the polymeric additive.

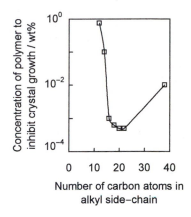

Fig. 9.1 The influence of polymer structure on its inhibitory properties (after Lewtas *et al.*, 1991).

A second example of this type of product is the use of inhibitors to reduce precipitation of insoluble inorganic species from water, so-called *scale inhibitors*. All natural water contains dissolved inorganic ions and the appearance of a furry coating of calcium carbonate on the heating element of a kettle is a common occurrence, particularly in areas of hard water. Domestic washing powders contain phosphates and sequestrants to prevent crystals depositing during the washing process. Industrial cooling water is treated with organic phosphonates and carboxylates to prevent scale build-up on heat exchanger surfaces. In off-shore oil recovery similar scale inhibitors are used to prevent blocking of pipelines and pores in oil-bearing rock.

In the context of off-shore oil recovery one particularly difficult scale to treat is barium sulphate, the most insoluble of the common inorganic scales. Boreholes are sunk into porous oil-bearing rock strata beneath the sea bed. Oil is recovered from this rock (which contains both oil and water within its pores) by pumping seawater into the bore hole and displacing the oil which is then pumped to the surface as an oil-in-water dispersion. Seawater contains sulphate ions while some oil-bearing rocks contain water that is rich in dissolved barium. In these cases, when the seawater is pumped into the pores barium sulphate is precipitated and the pores become blocked. This prevents further oil being recovered and can threaten the viability of the oil field.

Using the concept of tailor-made additives discussed in Chapter 5 it is possible to develop a design strategy to identify molecules which could be dissolved in the seawater and inhibit the crystallization of barium sulphate. Step one of this procedure involves identifying the crystal face(s) on which the additive is to act. This would normally be the fastest growing face. The second step is to visualize the molecular or ionic structure of the face in question.

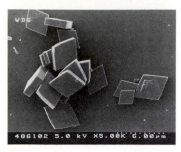

Fig. 9.2 The morphology of BaSO₄ crystals.

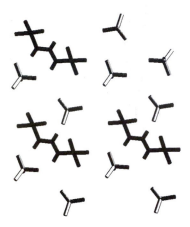

Fig. 9.3 Three diphosphonate molecules in the (011) surface of a barium sulphate crystal.

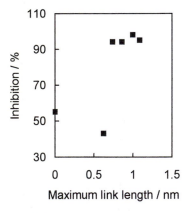

Fig. 9.4 The carbon chain length of inhibitor molecules has a pronounced effect on the precipitation of barium sulphate, as measured by the amount of Ba²⁺ left in solution after 4 hours (after Davey *et al.*, 1991).

Both these steps require knowledge of the crystal structure of the crystallizing material in question. The final step is to identify and test molecules that have appropriate functional and geometric features to bind strongly to the face and to act as an inhibitor.

Barium sulphate crystals form from seawater as (001) rhombic plates bounded by fast-growing {210} faces as shown in Fig. 9.2. Organic phosphonate anions have been tested as crystallization inhibitors since it would be anticipated that the divalent phophonate group might mimic the sulphate ion and allow such molecules to enter growing surfaces. The organic functionality would then disrupt growth. One particular molecular motif has been found to be highly effective—two phosphonate groups joined by a three-atom chain. An example of such a molecule is aminodimethylene diphosphonic acid. Figure 9.3 shows this molecule occupying two sulphate sites on a (011) face of barium sulphate. It is clear that the geometry of the molecule is particularly favourable in allowing both phosphonate groups simultaneously to access related sulphate sites. Energetic calculations confirm this and show that the three-atom chain then lays across the surface to form a steric and electrostatic barrier to further growth (Rohl *et al.*, 1996).

Using this model it is possible to extend the concept to the design of molecules that may be even more effective. For example, additive molecules that contain two of the diphosphonate motifs would be able simultaneously to access four sulphate sites and hence might be more active than molecules containing only one motif. Thus Fig. 9.3 shows the geometric juxtaposition of three additive molecules occupying adjacent surface sites. It also defines the geometry of such a molecule in that the link between two motifs must be at least a seven-carbon chain having a distance of 0.74 nm. To develop a *structure–property* relationship for this system, a series of such molecules having a range of link lengths from 0.60 to 1.20 nm was prepared and tested for their ability to inhibit the precipitation of barium sulphate from mixtures of barium chloride and sodium sulphate. Figure 9.4 shows the results of such studies. The sudden enhancement of inhibition when the link length reaches 0.74 nm is consistent with the principles on which the molecular design was based and is the basis for enhanced product activity (Davey *et al.*, 1991).

Two further examples of these principles of molecular additive design are to be found in preventing crystallization of dyes in dyesheets for thermal transfer printing (Davey *et al.*, 1997) and controlling the setting of cements (Coveney *et al.*, 1998).

In the former, dye dissolved in an amorphous polymer is coated onto a polymeric support sheet to create a ribbon of alternate red, blue and yellow dye. This dyesheet is then utilized to create coloured images from digitally recorded photographs. In order to achieve acceptable optical colour densities it is often necessary to have such high dye loadings that, over a period of months, crystals form in the dyesheet. This causes diffusion of the dye to the growing crystals and hence inhomogeneities in the dyesheet and in the subsequent colour prints. Incorporation of suitably designed additives into the formulation can prevent the nucleation and growth of crystals and hence extend the shelf life of the dyesheets.

The case of setting of cement is rather different. Here the setting process involves the dissolution and recrystallization of calcium silicates and aluminates with the mineral ettringite (calcium aluminium sulphate) being a key intermediate controlling the setting process. In order to extend setting times and so allow cements to be pumped as fluids to their point of application (in an off-shore oil well, for example), retardants must be designed to interact with and inhibit the formation of ettringite. As with barium sulphate, organic phosphonates are also highly active in this instance, particularly when designed with the appropriate geometry to match the crystal surfaces of ettringite. Most importantly from an application standpoint, setting is delayed without interfering with the subsequent mechanical properties of the cement.

9.5 Conclusions

This chapter has provided a brief overview of some aspects of crystals as they exist and transform in formulated products. Conceptually, this aspect of crystals and crystallization belongs more to the tradition of materials science than to that of physical chemistry or process technology. Materials science is concerned largely with the relationship between physical properties such as grain size, crystal structure and mechanical behaviour, and the micro and atomic level structure of materials such as metals, alloys, ceramics and polymers. There is now an increasing awareness that a similar approach would greatly benefit the area of product formulation.

References

Coveney P. V., Davey R. J., Griffen J. L. W. and Whiting A. (1998) *Chem. Commun.* 1467.
Davey R. J., Black S. N., Bromely L. A., Cottier D., Dobbs B. and Rout J. E. (1991) *Nature* **353** 549.
Davey R. J., Black S. N., Godwin A. D., Mackerron D., Maginn S. and Miller E. J. (1997) *J. Mater. Chem.* **7** 237.
Lewtas K., Tack R. D., Beiny D. H. M. and Mullin J. W. (1991) in *Advances in industrial crystallization*, J. Garside *et al.* (eds), p. 166, Butterworth-Heinemann, Oxford.
Rohl A. L., Gray D. H., Davey R. J. and Catlow C. R. A. (1996) *J. Amer. Chem. Soc.* **118** 642.

Further reading

Aulton M. E. (ed.) (1996) *Pharmaceutics: the science of dosage form design*, Churchill Livingstone, New York.

Index